Expanding the Circle

Scientific Breakthroughs that *Reinforce* Biblical Truths

And ye shall know the truth, and the truth shall make you free.

John 8:32

Speak to the earth, and it shall teach thee … that the hand of the Lord hath wrought this.

Job 12:8-9

Fifth Printing: August 2015

ISBN: 978-1-937735-74-6

Legends Library Publishing, Inc.
Rochester, New York

Inquiries to: info@legendslibrary.com
Phone: 877-222-1960
Website: www.LegendsLibrary.org

Cover design by Michael Hansen

Expanding the Circle

Scientific Breakthroughs that *Reinforce* Biblical Truths

Gary T. Wright

Legends Library

New York

Publisher's Note

For years I have been waiting for some brave soul to come along and declare the message found in this book. As a publisher, I have had a number of proposals cross my desk under the genre' of "reconciling science and the Bible." I have rejected every one of them. Why? Because they all essentially said the same thing, which was—*"Beginning with the unassailable foundation of science, I (the author) hereby set forth an interpretation of scripture that accommodates the hard facts of 21st century science."* In my way of thinking, these authors had it all backwards. They *assumed* that, since science has *obviously* been proven right, the Bible account must therefore be wrong, or should not be taken literally, and in fact requires some healthy reinterpretation. In effect, these "men of science" were raising their hands to "steady the ark" by correcting the 4000-year-old record, and this they do based upon their *superior* understanding of *established* science.

In the seventh chapter of Genesis we read of a universal flood that covered *"all the high hills, that were under the whole heaven… and the mountains were covered."* Now I know there are many who want to say that science has disproven the flood story and the account is therefore allegorical, or perhaps the flood was only "regional" in scope. I, however, tend to believe that the writer (Moses) knew exactly what *mountain* meant and what *covered* meant.

Despite holding a couple of advanced degrees, I do not claim to be a "scientist" in the classical sense; however, I have had occasion to study the nature of things over the course of my life, and it has often struck me that, with all the instruments of technology at our disposal today, we do not try to validate some of the *theories* which emerged from the minds of thinkers in the 1800's by subjecting them to actual tests measuring *empirical* evidence. Ironically, the scientific community suffers from an acute lack of curiosity when it comes to self-examination.

In today's intellectually enlightened world, we need honest brokers who are willing and courageous enough to ask the taboo questions of science. Questions such as: Why has contemporary science turned its back on God while the Fathers of Science believed in Him? Does our modern understanding of DNA support the theory of evolution proposed by Charles Darwin? Does the fossil record actually indicate that Darwinian evolution ever occurred anywhere?

In the author of this book, I believe we have found such an honest broker; one who is unafraid to examine the politically incorrect topics within the fields of earth science and biogeology. Wright has the courage to question modern science from a common sense viewpoint and allow the data to speak for itself.

Does this book claim to have all the answers? No, but it does represent a small spring bubbling out solid stone which will in turn feed other streams of truth and knowledge; which in time will grow into a swelling river of greater understanding. *Expanding the Circle* of human knowledge is a lifetime quest. May this work serve as a brick in the foundation of your quest for understanding and enlightenment.

Boyd J. Tuttle
Publisher
Legends Library Publishing Inc.
New York

CONTENTS

Figures

Prologue

Over the years since completing a Master of Science degree at the University of Colorado, the author has been watching with growing concern the role that mainstream contemporary science has undertaken to convince mankind that the entire universe and the miracle of life are merely the results of random events; thus implying that there is no Creator. Around the world, and especially in the United States of America, public education and academia both engage in nearly ubiquitous teaching and promotion of these secular concepts. The popular science television program NOVA, broadcast globally, provides powerful and credible reinforcement, as does much of the science content found on the Internet. As a result, virtually all Christian denominations are losing young adults in large numbers to the deadly clutches of atheism that logically follow such universal training.

In today's science community, any hypothesis that includes divine intervention is considered *pseudoscience* (without scientific foundation). No wonder that any new science which acknowledges the role of God as Creator is immediately rejected with contempt by a majority of the scientific community; by local, state, and national governments; and by the mainstream media.

Historically, science has not always taken such a stance, nor has it turned so defiantly away from God. Consider the following summary[1] of famous scientists (considered the fathers of science) who readily attested to God's role in the universe—

- **Nicholas Copernicus** (1473-1543)—developed the first mathematical model of the planets orbiting the sun. He referenced God in his works and saw no conflict between his hypotheses and the Bible.
- **Sir Francis Bacon** (1561-1627)—proposed the scientific method of inquiry based on experimentation and inductive reasoning. He believed that "depth in philosophy" brings men's minds to Providence and Deity.
- **Johannes Kepler** (1571-1630)—established the laws of planetary motion around the sun and introduced the concept of force in astronomy. Although a sincere Lutheran, he was allowed to teach at the Catholic university in Graz, Austria.
- **Galileo Galilei** (1564-1642)—pioneered the science of dynamics, and believed the Bible cannot err.
- **Rene Descartes** (1596-1650)—known as the father of both modern philosophy and analytical geometry, he, along with Sir Francis Bacon, developed the scientific methodology. Considered a genius, the Cartesian coordinate system was named after him. As a devout Catholic, he maintained a deep religious faith in God throughout his life.
- **Blaise Pascal** (1623-1662)—established the foundation for probability theory and developed the principles of air pressure and vacuums. After receiving a vision of God, he turned his studies from science to theology. His dying words were, *"May God never abandon me."*
- **Sir Isaac Newton** (1642-1727)—an undisputed genius, he invented calculus and developed classical physics, including the law of gravity. He was also deeply religious, although he did not believe in the doctrine of the Trinity. In his work

Principia, he wrote, *"The most beautiful system of the sun, planets, and comets, could only proceed from the counsel and dominion of an intelligent and powerful Being."* He also acknowledged that the Holy Ghost had led him to truth.

- **Robert Boyle** (1791-1867)—a founder of the Royal Society, he is known for Boyle's Law of gases and advancing chemistry. He also delighted in promoting Christianity and wrote against atheism.
- **Michael Faraday** (1791-1867)—one of the greatest scientists of the 19th century, his work on magnetism and electricity revolutionized physics. He was a devout Christian and belonged to a denomination that tried to emulate New Testament Christianity.
- **Gregor Mendel** (1822-1884)—an Austrian monk who laid the mathematical foundations of genetics while working in the garden of his monastery. He was eventually elected head of the monastery.
- **William Thomson Kelvin** (1824-1907)—foremost among the small group of British scientists who helped lay the foundations of modern physics. Many physicists say he had the most influence on 20th Century physics, and they classify him with Sir Isaac Newton and Albert Einstein in terms of his contributions to science. Like his fellow physicists George Gabriel Stokes and James Clerk Maxwell, Kelvin was deeply committed to Christianity during a time of religious apathy and anti-Christianity.
- **Max Planck** (1858-1947)—best known for quantum mechanics theory, he revolutionized our understanding of atomic and sub-atomic physics. Planck believed in an almighty, all-knowing, beneficent God, and felt that science and religion wage a *"tireless battle against skepticism and*

dogmatism, against unbelief and superstition" with the goal *"toward God!"*

- **Albert Einstein** (1879-1955)—probably the best known and most highly revered scientist of the 20th Century, he is associated with major revolutions in our thinking about time, gravity, and the conversion of matter into energy ($E=mc^2$). Although he did not believe in a personal God, Einstein recognized the impossibility of a non-created universe. A famous saying of his was, *"Science without religion is lame, religion without science is blind."* Einstein's famous response to the Heisenberg uncertainty principle was, *"God does not play dice,"* and to him it was a real statement about a God in whom he believed.

It is tragic that mainstream contemporary science has drifted so far from acknowledging all that God has done for mankind. Surely the Creator is offended over the direction modern science has taken—refusing to confess His hand in all things.

Many Christians find themselves in an uncomfortable dilemma when it comes to studying the tenets of contemporary science. They have been encouraged to embrace scientific truth, and yet the overriding (albeit indirect) theme of mainstream modern science is—*there is no God who created life and the universe.* How does one come to grips with such an apparent contradiction?

One of the author's former Sunday school teachers resolved this conundrum for himself by assuming that God guided the evolution of man, animals, and all living creatures on earth. Other Christians, including the author, reject this assumption. They do not believe that many of the tenets of contemporary science are true, and are looking for real scientific confirmation of their faith in God as The Creator, as set forth in the Bible.

The author's purpose in writing this book was to increase the reader's faith in God as The Creator by bringing to light new scientific breakthroughs that refute or challenge mainstream, Godless science. This new science, although considered pseudoscience by most entrenched scientists and therefore not accepted (as yet) by them, resolves the Christian dilemma discussed above and expands the revealed circle of truth which the reader undoubtedly cherishes.

The author recognizes that not all readers of this book will immediately adopt the conclusions of this new science, but the author also believes that those who are searching with an open mind and heart to strengthen their faith in God as The Creator will welcome the glorious truths revealed herein with joy and rejoicing.

Since this book discusses new science, a number of scientific terms are used with which the reader may be unfamiliar. The author provides a brief definition when each term is first introduced, and also provides a Glossary at the end of the book to which the reader may refer for a quick definition refresh.

All scriptures quoted in this book come from the King James Version of The Holy Bible.

Chapter One

Relevant Background Information

In order to better understand the new science that is introduced in this book, it will be helpful to first review some of the relevant truths that have already been revealed through the Bible.

Laws and Conditions of Mortal Life on Earth

Regarding the creation of the earth and its inhabitants, the Lord revealed the following—

And God said, Let the earth bring forth grass, the herb yielding seed, and the fruit tree yielding fruit after his kind, whose seed is in itself, upon the earth: and it was so. And the earth brought forth grass, and herb yielding seed after his kind, and the tree yielding fruit, whose seed was in itself, after his kind: and God saw that it was good. (Genesis 1:11-12)

And God created great whales, and every living creature that moveth, which the waters brought forth abundantly, after their kind, and every winged fowl after his kind: and God saw that it was good. (Genesis 1:21)

And God said, Let the earth bring forth the living creature after his kind, cattle, and creeping thing, and beast of the earth after his kind: and it was so. And God made the beast of the earth after his kind, and cattle after their kind, and every thing that creepeth upon the earth after his kind: and God saw that it was good. (Genesis 1:24-25)

So God created man in his own image, in the image of God created he him; male and female created he them. (Genesis 1:27)

From these scriptures we learn about a fundamental restriction the Lord imposed upon all living things when the earth was created—that they bring forth after their own *kind*. This restriction applies to mankind as well, and has important implications for evolution that will be discussed in Chapters 2 and 4.

After Adam and Eve partook of the forbidden fruit in the Garden of Eden, God implemented significant changes to the earth and all of its inhabitants. Following Adam's transgression, God spoke to him and pronounced—

Because thou hast hearkened unto the voice of thy wife, and hast eaten of the tree, of which I commanded thee, saying, Thou shalt not eat of it: cursed is the ground for thy sake; in sorrow shalt thou eat of it all the days of thy life; Thorns also and thistles shall it bring forth to thee; and thou shalt eat the herb of the field; In the sweat of thy face shalt thou eat bread, till thou return unto the ground; for out of it wast thou taken: for dust thou art, and unto dust shalt thou return. (Genesis 3:17-19)

In effect, the Lord established here some fundamental laws and conditions that govern mortal life on earth, all but one of which differ from laws and conditions that were in place on earth and in the Garden of Eden before the Fall of Adam and Eve occurred. These laws and conditions of mortality (life on earth), along with some resulting consequences, include the following—

1. All living things will eventually die; therefore, they have a beginning and an end.
2. All living things that reproduce can only bring forth after their own kind. This law was in place before the Fall and is still in effect today.

3. Natural processes always proceed in the direction of less order (or more randomness); also, any effort to increase order requires work, or the expenditure of energy. A resulting consequence of this condition is that *everything* in mortality naturally degenerates; hence the cause of death.

A consequence of the first condition listed above is that the human mind cannot fully comprehend eternity. The second condition identified above has a fundamental impact on the scope of evolution, as is discussed in Chapters 2 and 4.

Modern science has formally restated the third condition of mortality listed above as the Second Law of Thermodynamics. This law introduces the concept of *entropy*, a somewhat difficult concept to grasp, but one that is key to understanding the new science covered in Chapter 2. In essence, the Second Law of Thermodynamics states that a spontaneous process involving a closed system will always maximize entropy, or progress toward thermodynamic equilibrium, and thus disperse energy. On a microscopic level, entropy is defined as the measure of randomness (or disorder) of the molecules within a thermodynamic system.

To illustrate entropy, consider an ice cube that is placed on a table in a room that is at a comfortable temperature. What will *always* happen to the ice cube? Of course, it will melt. Why?

1. Because water molecules in an ice cube are well-ordered in a solid, crystalline structure and require a temperature below 32° F (0° C) to maintain their crystalline structure. Since the room temperature is well above 32° F (0° C), the Second Law of Thermodynamics requires that entropy—the measure of disorder or randomness of the molecules in the crystalline structure—increase. This happens as the ice cube melts. The molecules in the melted water are no longer in a crystalline structure; rather, they have no regular arrangement and thus

have more randomness or less order compared to their previous crystalline structure. Thus entropy has increased in the melted water and the Second Law of Thermodynamics prevails.

2. Heat is transferred from the table and the warm air in the room to the ice cube. Thus heat energy is dispersed from the air and table into the ice cube, raising the temperature of the crystalline structure and melting the ice into water. The closed system in this case consists of the ice cube, the table, and the air in the room. Thus energy has been dispersed in the closed system and the Second Law of Thermodynamics prevails.

3. Energy had previously been expended to freeze the ice cube and maintain its crystalline structure. That energy was in the form of electricity, which powered the freezer from which the ice cube was taken. When the ice cube was removed from the freezer and placed on the table, it became part of a different system. Since energy is no longer being expended to keep it frozen on the table, the Second Law of Thermodynamics once again prevails. This explanation illustrates why work is required to bring about more order, or to make things better or more refined, in mortality.

The melting ice cube illustrates why physical things naturally degenerate in mortality—they continually regress toward disorder unless work is applied to reverse the regression. Not surprisingly, a second definition of entropy is the process of degradation, or degeneration, that naturally takes place with all things physical on earth. The aging process of all living organisms is a good example of this definition of entropy.

The concept of entropy also highlights the phenomenon that energy is dispersed in a natural process. From where does the energy come in the first place? After energy is dispersed or expended, it is no longer available. Thus energy must be supplied from *somewhere* before a natural process can occur.

This concept may be more easily understood by considering a wind-up clock—nothing happens until the clock is wound; it then proceeds to keep time as the spring mechanism winds down. When the mechanism is totally unwound, it stops functioning. The clock will only keep time once again if someone rewinds it.

Such is mortality. Expenditure of energy is required for everything that occurs. As you watch in wonder as majestic, billowy white clouds drift silently and effortlessly across the deep blue sky, do you ever wonder from where the energy came to move them? Who or what wound up the world in the first place? Who or what keeps it wound up? Christians know the answer—God.

When the author studied thermodynamics, the writers of his textbook[1] (Gordon J. Van Wylen and Richard E. Sonntag) made the following comment about the Second Law of Thermodynamics—

"... the authors see the second law of thermodynamics as man's description of the prior and continuing work of a creator, who also holds the answer to the future destiny of man and the universe."

It is refreshing to see some scientists acknowledge God in their work!

The effects of entropy, as they apply to evolution, will be discussed in more detail in Chapter 2.

The Flood

A significant area of disagreement between the scriptures and mainstream contemporary science is the extent of the Flood God sent to cleanse the earth. Here is what Genesis records about the Flood—

In the six hundredth year of Noah's life, in the second month, the seventeenth day of the month, the same day were all the fountains of the great deep broken up, and the windows of heaven were opened. And the rain was upon the earth forty days and forty nights. (Genesis 7:11-12)

And the flood was forty days upon the earth; and the waters increased, and bare up the ark, and it was lift up above the earth. And the waters prevailed, and were increased greatly upon the earth; and the ark went upon the face of the waters. And the waters prevailed exceedingly upon the earth; and all the high hills, that were under the whole heaven, were covered. Fifteen cubits upward did the waters prevail; and the mountains were covered. And all flesh died that moved upon the earth, both of fowl, and of cattle, and of beast, and of every creeping thing that creepeth upon the earth, and every man. All in whose nostrils was the breath of life, of all that was in the dry land, died. And every living substance was destroyed which was upon the face of the ground, both man, and cattle, and the creeping things, and the fowl of the heaven; and they were destroyed from the earth: and Noah only remained alive, and they that were with him in the ark. And the waters prevailed upon the earth an hundred and fifty days. (Genesis 7:17-24)

And God remembered Noah, and every living thing, and all the cattle that was with him in the ark: and God made a wind to pass over the earth, and the waters assuaged. The fountains also of the deep and the windows of heaven were stopped, and the rain from heaven was restrained; And the waters returned from off the earth continually: and after the end of the hundred and fifty days the waters were abated. And the ark rested in the seventh month, on the seventeenth day of the month, upon the mountains of Ararat. And the waters decreased continually until the tenth month: in the tenth month, on the first day of the month, were the tops of the mountains seen. (Genesis 8:1-5)

And it came to pass in the six hundredth and first year, in the first month, the first day of the month, the waters were dried up from off the earth: and Noah removed the covering of the ark, and looked, and, behold, the face of the ground was dry. And in the second month, on the seven and twentieth day of the month, was the earth dried. (Genesis 8:13-14)

Christians have been taught that the Flood was truly global. The scriptures above certainly indicate that the entire world was submerged for a period of time measured in months. It took more than seven months from the beginning of the Flood until the tops of the mountains were seen, and Noah did not depart from the ark until a complete year and ten days had passed. Some Bible scholars claim that the Flood lasted 370 days—150 days of flooding followed by 220 days of the waters receding back to the ocean. However, Genesis states that God told Noah, and everything with him, to go forth from the ark 375 days after the flood started.

The scientific community does not believe that the Flood was global; rather, it is thought to have been nothing more than a regional flood in the Mesopotamian Valley—if it occurred at all. Most scientists believe the Bible's account of the Flood is a myth.

Whether or not the Flood was global has enormous consequences for the science of geology. This topic will be discussed in more detail in Chapters 3 and 4.

Laws and Conditions of Eternity

One important truth revealed in the Old Testament is—

For my thoughts are not your thoughts, neither are your ways my ways, saith the Lord. For as the heavens are higher than the earth, so are my ways higher than your ways, and my thoughts than your thoughts. (Isaiah 55:8-9)

From this verse one may infer that there are different laws and conditions (i.e., "ways") that apply to the heavens and the mortal earth. Consider the following scripture—

But is now made manifest by the appearing of our Saviour Jesus Christ, who hath abolished death, and hath brought life and immortality to light through the gospel: (2 Timothy 1:10)

Jesus Christ abolished death through the resurrection—which replaces mortality with immortality. All who have lived on this earth, or who will yet live on earth, will be resurrected with an immortal body!

For as in Adam all die, even so in Christ shall all be made alive. (1 Corinthians 15:22)

How can this be? The Savior explained it this way—

In my Father's house are many mansions: if it were not so, I would have told you. I go to prepare a place for you. And if I go and prepare a place for you, I will come again, and receive you unto myself; that where I am, there ye may be also. (John 14:2)

Christ made it clear in this scripture that His faithful followers will dwell with Him in His Father's house after they are resurrected. This promise implies that mortality will no longer apply after the resurrection; immortality will reign instead—confirming that there are truly different laws and conditions which govern life in different parts of the universe. How wonderful and glorious to contemplate that the laws and conditions of mortality are merely a temporary step to eternal life!

That as sin hath reigned unto death, even so might grace reign through righteousness unto eternal life by Jesus Christ our Lord. (Romans 5:21)

No mainstream scientist has any inkling about the way God's universe actually functions, as revealed in these enlightening verses. The only paradigm for the visible universe such a scientist can imagine is limited by the laws and conditions of mortality, presented above, that determine his/her own existence.

As one prominent geomorphology scientist, Dr. Arthur N. Strahler, wrote[2] (highlights are the author's, not Strahler's)—

"Uniformitarianism should be viewed as antithetic to any and all theories invoking supernatural forces for the origins of things geologic or biologic or of any other natural-science phenomena. Uniformitarianism includes the assumption that ***laws of physics and chemistry as presently understood have applied to all past time and will apply to all future time****; they are assumed valid unless compelling independent evidence should be found to suggest that they may have differed in quality or quantity at other points in time or space. In using that assumption,* ***all science operates today under the uniformitarian principle****. In cosmology, the study of the structure and evolution of the universe,* ***it is assumed that the laws of physics are similar throughout the entire universe****."*

The fact that different laws and conditions apply to mortality and immortality is unknown to mainstream contemporary science; therefore, Christians should keep in mind that the theories and hypotheses of modern science do not reflect the realities of God's universe that are revealed in the Bible.

CHAPTER TWO

Evolution and the Primary Axiom

Evolution is essential to maintain life on earth.

However, this statement of truth requires qualification.

Two types of evolution will be discussed in this chapter. The first is adaptive evolution (as opposed to forward evolution which is discussed below). Adaptive evolution utilizes the adaptive genetic capabilities that were given to all living things when they were created. Such capabilities allow organisms to evolve within a *kind* in a way that allows them to better survive in a given habitat.

Remember that *kind* is the term used in Genesis to explain the reproductive restriction imposed upon all living things—that they bring forth after their own *kind*. As a biblical classification, kind corresponds to our modern biological taxonomic ranks of *genus* and *family*. The rank of genus is placed above *species*, and the rank of family is placed above genus. Kind encompasses both.

Liger
(courtesy Ali West, Wikimedia Commons)

The reason that genus ranks above species is that many related species can interbreed and produce fertile offspring. The liger shown here is an example of such *hybridization*—it is a lion/tiger hybrid. Besides big cats, other genera that hybridize include dogs, bears, deer, cattle, rodents, birds, reptiles, etc.

After the Great Flood, the only *kinds* of organisms that had survived the catastrophic deluge included those saved in Noah's ark and those already adapted to the waters of the Flood itself (such as fish and whales). Although the ark is thought to have had a volumetric capacity of more than 1.5 million cubic feet[1] (42,475 cubic m), this large amount of passenger and food storage space could not hold a male/female pair from all the non-aquatic species that are found on earth today. So how did all of these species evolve?

The answer is adaptive evolution. This is best illustrated with an example. There was probably only a single male/female pair of big cats on the ark. Today, the big cat genus *Panthera* comprises lions, tigers, leopards, and jaguars.

Lion
courtesy Kevin Pluck, Wikimedia Commons

Tiger
courtesy J. Patrick Fischer, Wikimedia Commons

Leopard
courtesy P. Lindgren, Wikimedia Commons

Jaguar
courtesy Wikimedia Commons

Moving up a level to the *Felidae* family, cougars, cheetahs, snow leopards, and clouded leopards are also included in the family of big cats that roam their respective habitats around the globe.

Cougar
courtesy Wikimedia Commons

Cheetah
courtesy Guarav P., Wikimedia Commons

Snow Leopard
courtesy Aaron Logan, Wikimedia Commons

Clouded Leopard
courtesy Nancy Vandermey from the Feline Conservation Center, Rosamond, CA, Wikimedia Commons

Notice the fascinating variety of fur coats with which these eight gorgeous species are endowed; and yet, they are clearly all big cats from the same family (or kind) that left the ark some 4500 years ago. This striking example of adaptive evolution illustrates how living organisms can evolve *within a kind* to better survive in a particular habitat.

Melanistic Jaguar
courtesy U.S. Fish and Wildlife Service, Wikimedia Commons (in the public domain)

Typical Jaguar
courtesy Cburnett at the Milwaukee County Zoological Gardens,

A variation of adaptive evolution is known as *polymorphism*, which occurs when different *phenotypes* (sets of observable traits) exist within the same population of a species. For example, the melanistic (black) jaguar shown here (also known as a black panther) represents about 6% of the South American population of jaguars. Melanistic jaguars freely interbreed with typical jaguars in South America since they are of the same species.

Common Pheasants
courtesy ChrisO at the English language Wikipedia, Wikimedia Commons

Perhaps a more familiar type of polymorphism is *sexual dimorphism,* which differentiates males from females within a species in traits such as shape, structure, color, pattern, size, ornamentation, and behavior. Notice the dramatic difference in both color and size between common pheasant sexes shown here (the female is on the left). This type of polymorphism is especially prevalent among birds.

Without adaptive evolution, a population, or even an entire species, could become extinct if its habitat were to change over time, or if the population were forced to migrate to and remain in a new habitat. Thus, adaptive evolution is God's way of ensuring survival of his living creations in a dynamic and hostile mortal world. Adaptive evolution was especially important after the animals on Noah's ark left their wooden, seafaring home and began to repopulate the earth. Adaptive evolution must have been operative in the progeny of Shem, Ham, and Japheth as well, as evidenced by the different physical characteristics found in humans around the world.

A controversial scientific explanation of adaptive evolution was published by Motoo Kimura in 1983. It is known as "The Neutral Theory of Molecular Evolution." This theory proposes that, at the molecular level, adaptive evolutionary changes that spread across a species' population are primarily brought about by genetic mutations that are nearly-neutral (or nearly invisible) with respect to natural selection. Therefore, the fate of the mutations is mainly determined by random genetic drift in the population rather than by natural selection. Random genetic drift will be covered in more detail later in this chapter.

Kimura's theory is controversial because it can be viewed as being hostile to Charles Darwin's claim of evolution by natural selection. However, Kimura said[2] of his theory—

"It is not antagonistic to the cherished view that evolution of form and function is guided by Darwinian selection, but it brings out another facet of the evolutionary process by emphasizing the much greater role of mutation pressure and random drift."

Kimura's theory will be discussed in more detail later in this chapter.

The type of evolution Darwin believed natural selection can bring about may be thought of as forward (or progressive) evolution. This type of evolution supposedly results in the gradual appearance on earth of new species that are *more complex, or a higher order,* than the species from which they evolved. While adaptive evolution only brings about *changes within a kind* (genus or family), forward evolution is believed by mainstream science to bring about entirely new species that are *genetically superior* to the species from which they evolved, and hence are *new kinds that did not exist previously*. The genetic superiority supposedly results from the creation of new and additional genetic (biological) information (i.e., DNA). Via forward evolution, man is thought to have ultimately descended from the most primitive form of life, which somehow gained its life from self-replicating but nonliving molecules.

Of course, natural processes that could result in life coming from nonliving matter, or could result in a genome gaining additional biological information that significantly improves the genome, would require entropy to decrease (or order to increase), and thus these processes would break the Second Law of Thermodynamics. More about this topic will be covered below. (Note: a *genome* is the complete set of an organism's hereditary information—in other words, its book of life. A genome is the sum of an organism's genetic parts, including all of its genes and chromosomes. Genome may also refer to the complete set of an entire species' hereditary information, such as the book of life of the human race.)

The theory of forward evolution involves random genetic mutations and extremely long periods of time for natural selection to do its work. As the popular science author Carl Sagan explained[3] it—

"The secrets of evolution are death and time—the deaths of enormous numbers of life forms that were imperfectly adapted to the environment; and time for a long succession of small mutations that were by accident adaptive, time for the slow accumulation of patterns of favorable mutations."

Thus forward evolution is positioned by its champions to be the result of a pointless, unguided, amoral natural process. There is no room in this characterization for God to guide forward evolution; indeed, there is no room in this characterization for God to even exist.

The fundamental basis upon which forward evolution is built is labeled by some scientists "The Primary Axiom." An axiom is defined as an established premise, principle, or rule, or a self-evident truth. *The Primary Axiom is that man is merely the product of random mutations and natural selection.*[4]

Within the community of mainstream scientists, the Primary Axiom is duly and even blindly accepted as a self-evident truth. However, to the author and a growing number of others, it is *not* a self-evident truth; rather, it is a false premise that can be exposed by critical scientific analyses and common sense. In fact, the Primary Axiom is the antithesis of the definition of an axiom.

Christians should understand that mainstream contemporary science is "religiously"—even fanatically—committed to forward evolution. Note the words[6] of Dr. Michael Ruse, who is widely considered the leading spokesman for evolution today—

"Evolution is promoted by its practitioners as more than mere science. Evolution is promulgated as an ideology, a secular religion—a full-fledged alternative to Christianity, with meaning and morality. I am an ardent evolutionist and an ex-Christian, but I must admit that in this one complaint ... the literalists are absolutely right. Evolution is a religion. This was true of evolution in the beginning, and it is true of evolution still today."

Even more startling, consider the words[7] of Dr. Richard Lewontin, another leading evolutionist (emphases are the author's, not Lewontin's)—

"We take the side of science in spite of the ***patent absurdity of some of its constructs****, in spite of its failure to fulfill many of its extravagant promises of health and life, in spite of the* ***tolerance of the scientific community for unsubstantiated just-so stories****, because we have a prior commitment to materialism ... We are forced by our* ***a priori*** *adherence to material causes to create an apparatus of investigation and a set of concepts that produce material explanations, no matter how counterintuitive, no matter how mystifying to the uninitiated. Moreover,* ***that materialism is absolute, for we cannot allow a Divine Foot in the door****."*

The truthfulness of these two quotations will become evident as more new science is presented and discussed in this book.

Evolutionists claim that the history of forward evolution dates back to the 1500s when Leonardo da Vinci proposed that fossils are the remains of animals that were buried in soft sand and mud. It will be shown in Chapters 3 and 4 that he was more correct than evolutionists would like to admit today.

Charles Darwin did not make his mark on forward evolution until 1859, when his major work, *On the Origin of Species by Means of Natural Selection,* was published. This work immediately gained acceptance by many scientists and even much of the public, effectively launching the new "religion."

Unknown to many, the road that forward evolution has taken since Darwin has not been smooth; rather, it has required "absurd constructs" and "unsubstantiated just-so stories" to keep it moving over the bumps that have been encountered as new science has emerged over the years. Some of these significant bumps are summarized below. They provide a solid foundation for proving that the Primary Axiom is false, as will be shown later in this chapter.

Haldane's Dilemma

One such bump came in 1957 when Jack Haldane, a British-born geneticist and evolutionary biologist, had a paper[8] published titled "The Cost of Natural Selection." Haldane was a staunch Marxist and is considered one of the three founders of population genetics. He was especially strong in the mathematical theory of genetics. (Note: population genetics is a vital element of modern evolutionary science). In this paper, Haldane made the following statements (emphases are the author's, not Haldane's)—

"It is well known that breeders find difficulty in selecting simultaneously for all the qualities desired in a stock ... in this paper I shall try to make quantitative the fairly obvious statement that ***natural selection cannot occur with great intensity for a number of characters at once*** *..."*

"... the number of deaths needed to secure the substitution by natural selection of one gene ... is about 30 times the number of organisms in a generation ... ***the mean time taken for each gene substitution is about 300 generations****."*

"If two species differ at 1,000 loci, and the mean rate of gene substitution, as has been suggested, is one per 300 generations, ***it will take at least 300,000 generations (6 million years)*** *..."*

"Even the geological time scale is too short for such processes to go on in respect to thousands of loci ... can this slowness be avoided by selecting several genes at a time? I doubt it ..."

Haldane's statement above that, *"natural selection cannot occur with great intensity for a number of characters at once"* recognizes a serious limitation of natural selection that will be discussed in greater detail later in this chapter. Haldane also states above that, for man, it would take six million years for a mere one thousand mutations to become a permanent part of the human genome through natural selection (assuming twenty years between generations). Six million years is the amount of time it supposedly took for mankind to evolve from chimpanzees! Is the genetic difference between man and chimpanzee only one thousand mutations? Obviously not!

Haldane could not know back in the 1950s that the number of genetic units in the human genome, which are *all* potential mutation sites, is actually *three billion,* and that at least *one million new mutations are entering the human genome every generation!*[9] The sheer size of the human genome, along with its mutation rate—as we understand them today—confirm the truth of Haldane's statement above that, ***"Even the geological time scale is too short for such processes to go on ..."*** Thus the Haldane Dilemma.

In 1993, Walter ReMine, an electrical engineer with a strong mathematical background, published *The Biotic Message,* a book that independently confirms Haldane's Dilemma and provides an alternative theory to forward evolution known as the message theory.

To date, evolution scientists have not adequately resolved Haldane's Dilemma. Of course they offer various arguments why it is not a problem, but for those who are not "converted" to the Primary Axiom, Haldane's Dilemma remains an open challenge to the theory of forward evolution.

Kimura's Quandary

Motoo Kimura was an influential theoretical population geneticist who was a fellow of the Royal Society and was awarded the Darwin Medal in 1992. As such, he was committed to the Primary Axiom. Kimura was mentioned previously for his scientific explanation of adaptive evolution in the book "The Neutral Theory of Molecular Evolution." In this book, that was first published[10] in 1983, Kimura made the following statements (emphases are the author's, not Kimura's)—

"This formula shows that ***as compared to Haldane's formula the cost is larger by about 2*** *... under the assumption that the majority of mutation substitutions at the molecular level are carried out by positive selection ...* ***to maintain the same population number and still carry out mutant substitutions ... each parent must leave ... 3.27 million offspring to survive and reproduce.*** *This was the main argument I used when I presented the neutral mutation-random drift hypothesis of molecular evolution ..."*

Here Kimura acknowledges that Haldane was conceptually right, although he doubled Haldane's cost of natural selection. Like Haldane, Kimura understood that natural selection must occur very slowly and that it can only affect a small number of mutations at any one time. The evolutionists of Kimura's time were expecting far too much of natural selection—leading to an absurd "cost" of more than *three million reproducing offspring required for each parent!* Because Kimura realized this calculated cost of natural selection was too great to be taken seriously, he proposed his neutral theory in order to prop up the Primary Axiom.

Neutral mutations are so named because they have essentially a nearly-neutral (or nearly-zero) effect on an organism's fitness, and therefore are not selectable by natural selection. Ironically, Kimura's neutral theory of molecular evolution insists that *most* mutations are unselectable—so most genetic information must be irrelevant and, therefore, *most evolution must not be the result of natural selection!*

Thus Kimura's neutral theory provides strong support *for adaptive* evolution—while simultaneously offering a rational argument *against forward* evolution. Kimura's nearly-neutral mutations will be discussed in more detail later in this chapter.

Muller's Fear

An atheist from his high school days, Hermann Muller was a geneticist who was awarded a Nobel Prize in 1946 for his discovery that mutations can be induced by x-rays. He was also a proponent of *eugenics*—the practice of improving the quality of the human race by promoting increased reproduction of people with desired traits and less reproduction of people with undesired traits. Because of Muller's work with mutations, he came to the realization that **the human genome is deteriorating!** In 1950, the *American Journal of Human Genetics* published a paper[11] by Muller titled, "Our Load of Mutations." Some of his statements in this article are captured here (note: **u** stands for mutation rate; also, emphases are the author's, not Muller's)—

*"... **it is very difficult to estimate the rate of the degenerative genetic process** ... it is evident that the natural rate of mutation of man is so high, and his natural rate of reproduction so low, that not a great deal of margin is left for selection... if u has the minimal value of 0.1 ... an average reproductive rate of 2.4 children per individual would be necessary... **without taking any account whatever of all the deaths and failures to reproduce for non-genetic causes... it becomes perfectly evident that the present number of children per***

couple cannot be great enough to allow selection to keep pace with a mutation rate of 0.1 ... if, to make matters worse, u should be anything like as high as 0.5..., our present reproductive practices would be utterly out of line with human requirements."

*"... **the present genetic load is a serious one... increasing it by only 25% ... would be a matter of grave consequence, while its doubling would be calamitous** ... **if u should rise above 0.5, the amount of selective elimination required ... would, as we have seen, be greater than the rate of effective reproduction of even primitive man would have allowed ... genetic composition would deteriorate continuously, while the population would meanwhile diminish in number, all the way to the point of disappearance.**"*

*"... in many civilized nations, the birth rate is held down to an average of not much more than two per family, **the upper mutation rate, that beyond which equilibrium is impossible, must be much lower than 0.5 and, as we have seen, perhaps lower than 0.1, even if selection were to be given full scope.**"*

"If we now postulated that the conditions of raised mutation rate, low birth rate... it would be very problematic whether or not this decline would eventually be arrested."

A deteriorating genome cannot support forward evolution, which requires the genome to be progressing, not deteriorating! Muller's bump in the road of forward evolution is huge—it is actually the crux of this chapter, and the basis for determining if the Primary Axiom is true or false.

Muller clearly feared that the mutation rate of the human genome is so high that degeneration of the genome cannot be stopped with today's fertility rates, even if forced selection (i.e., eugenics) were to be implemented. As will be discussed in detail later, **modern science has learned that the real mutation rate (u) is somewhere between 100 and 300!** Knowing this, Muller's statement above—*"if u should rise above 0.5, the amount of selective* *elimination required ... would, as we have seen, be greater than the rate of effective reproduction of even primitive man would have allowed... genetic composition would deteriorate continuously, while the population would meanwhile diminish in number, all the way to the point of disappearance,"*—provides scientific proof that the Primary Axiom is false, and that forward evolution could not have been responsible for the appearance of man on the earth.

Because of Muller's Fear, he spent much of his life educating the public on the dangers of radioactive fallout coming from nuclear testing and nuclear war.

Muller's Ratchet

One of Muller's later papers,[12] published in 1964, explained his thoughts about nearly-neutral mutations—

"There comes a level of advantage, however, that is too small to be effectively seized upon by selection, its voice being lost in the noise, so to speak."

This statement introduces the concept of *biological noise* diminishing the effectiveness of natural selection. Examples of such noise include (*i*) an organism's environment and its interaction with that environment, (*ii*) gene changes that are not DNA changes (this is known as *epigenetics*), (*iii*) mutated genes interacting with other modifier genes (this is known as *epistasis*), and (*iv*) interaction of the members of a single gene pair (this is known as *dominance*).

Genetic scientists have estimated that about 50% of all observable (i.e., phenotypic) variation within a population is caused exclusively by environment variations. Another 25% results from the way individual organisms in the population interact with their own local environment. 24% of observable variation comes from genetic changes that are not heritable, and therefore cannot be selected by natural selection. Thus, only about

1% of observable variation within a population is *heritable* (able to be inherited), and therefore available for selection.[13]

In his neutral theory of molecular evolution, Kimura indicated[14] that the heritability of an organism's *fitness* (ability to survive and reproduce) is only about 0.4%, and therefore only 0.4% of observable variation for fitness is selectable, or available for natural selection. 0.4% corresponds to a signal-to-noise ratio of 1:250. This ratio quantifies why natural selection for total observable fitness results in a nearly-zero gain toward forward evolution.[15]

Thus Muller raised another concern about the potential reach of natural selection.

Although he was active in warning of the dangers of radiation, Muller was very careful to not make statements that might imply a lack of confidence in the Primary Axiom. However, the paper referenced above included a statement that, as explained below, *does* argue against the Axiom—

"... an asexual population incorporates a kind of ratchet mechanism, such that ... lines become more heavily loaded with mutation."

Asexual organisms typically include single cell bacteria and certain types of plants and fungi that reproduce asexually. Asexual reproduction occurs when offspring have only a single parent and thus inherit the genes of that parent only.

A *ratchet* is a device that allows motion in one direction while preventing motion in the opposite direction.

In his paper, Muller argued that his "ratchet" only applies to asexual populations because, once a mutation becomes *fixed* (permanent) in an asexual genome, it can never be removed from that genome. Over time, the information in the genome will degenerate since each new generation will have more mutations than the previous generation. Thus, only a unidirectional (i.e., ratcheted) process is possible for an asexual population.

Ironically, Muller's introduction of biological noise as a limiting factor for natural selection in this same paper requires that his "ratchet" apply to sexual populations as well! Because of biological noise, a nontrivial percentage of *deleterious* (damaging or harmful) mutations will escape natural selection and become fixed, or permanent, in the genome of a sexual species. As shall be shown later, deleterious mutations always vastly outnumber beneficial mutations. Natural selection cannot discern between good and bad mutations because they exist together in large linkage blocks and they all function together. Therefore, each linkage block in the genome of a sexual species will degenerate individually because of Muller's Ratchet.

Neel's Realization

James Neel was an American geneticist who established the first university department of genetics in the United States (at the University of Michigan). He was also instrumental in establishing human genetics as a research field in the United States. In a paper[16] which Neel authored (along with six other Japanese contributors) that was published in the *Proceedings of the National Academy of Science of the United States of America* in 1986, the following statements were made (emphases are the author's, not Neel's)—

"If this approximation is correct, then gamete rates for "point" mutations become of the order of 30 per generation… ***The implications of mutations of this magnitude for population genetics and evolutionary theory are profound.*** *The first response of many population geneticists is to suggest that most of these occur in "silent" DNA and are of no real biological significance. Unfortunately for that line of reasoning …* ***The amount of silent DNA is steadily***

shrinking. The question of how our species accommodates such mutation rates is central to evolutionary thought."

A *gamete* is a reproductive cell—either a sperm or an ovum cell in humans. Gamete mutations are particularly important from an evolutionary point of view because the mutations always get passed along to offspring if the mutated gametes are fertilized. Neel's work showed that gamete mutation rates are so high they cannot be ignored by evolutionists.

At the time (1986), many evolutionists were insisting that the vast majority of DNA in the human genome is "junk" or "silent" DNA, meaning that it has no known biological function and therefore does not participate in forward evolution. Neel proactively countered their potential attempt to discredit his high mutation rate by pointing out that, *"The amount of silent DNA is steadily shrinking."*

He was right. In September 2012, the journal *Science* reported that the Encyclopedia of DNA Elements (ENCODE) Project, which is a research program supported by the National Human Genome Research Institute, *"found that 80% of the human genome serves some purpose, biochemically speaking,"* and therefore is functional after all.[17] Of course, this finding has been criticized by evolutionists who realize what it could mean to the theory of forward evolution. Nevertheless, this *Science* article also reported that—

"This week, 30 research papers, including six in Nature *and additional papers published online by* Science, *sound the death knell for the idea that our DNA is mostly littered with useless bases."*

Neel's Realization is yet another bump in the road that forward evolution has encountered since Darwin. More about junk DNA will be covered later in this chapter.

Kondrashov's Question

Originally from the Soviet Union, Alexey Kondrashov has been working in the United States since the 1990s. He is currently a professor at the University of Michigan and is known for his work on evaluating genetic mutation rates. The title of his paper[18] published in the *Journal of Theoretical Biology* in 1995 was "Contamination of the genome by very slightly deleterious mutations: **why have we not died 100 times over?**" In this paper, Kondrashov makes some relevant statements about degeneration of the human genome (emphases are the author's, not Kondrashov's)—

*"**Tachida (1990) concluded that VSDMs (very slightly deleterious mutations)** impairing only one function – its interaction with nucleosomes – **may lead to too high a mutation load**..."*

*"Lande (1994) and Lynch et al (1994) ...concluded... **VSDMs can rapidly drive the population to extinction ...**"*

*"**Simultaneous selection against many mutations can lead to a further decline of** N_e* (author's note: N_e is the effective population size) ***and facilitate extinction** (Li 1987; Lynch at al 1984)."*

*"**I interpret the results in terms of the whole genome and show, in agreement with Tachida (1990), that VSDMs can cause too high a mutation load**, even when $N_e = 10^6$-10^7 ... conditions under which the load may be paradoxically high are quite realistic..."*

"... the load can become excessive even when U<1 (author's note: U is deleterious, or harmful, mutations per person per generation) *... as my analysis suggests—**contamination by VSDMs implies an excessive load, leading to stochastic mutation load paradox.**"*

*"... **because the stochastic mutation load paradox appears real – it requires a resolution.**"*

*"**Chetverikov (1926) assumed the mutational contamination of a species increases with time, leading perhaps to its eventual extinction.**"*

*"... **accumulation of VSDMs in a lineage ... acts like a time bomb ... the existence of vertebrate lineages ... should be limited to 10^6-10^7 generations.**"*

In the quotations above, Kondrashov refers to five other scientists, besides himself, who have come to the realization that the human genome is degenerating, and that such deterioration will eventually lead to extinction of the race. Kondrashov's Question, *"Why have we not died 100 times over?"* and his analogy *"acts like a time bomb"* dramatically confirm the reality of Muller's Fear. Clearly, this persistent bump in the road of forward evolution—that the human genome is deteriorating rather than progressing and advancing—is becoming larger.

Kondrashov's Numbers

Near the end of 2002, Dr. Kondrashov wrote another paper[19] that was published in the journal *Human Mutation*. This article stated—

*"... the total number of new mutations per diploid human genome per generation is about 100 ... at least 10% of these are deleterious ... analysis of human variability suggests that **a normal person carries thousands of deleterious alleles** ..."*

An *allele* is a member of a gene pair that resides on a chromosome and is the mechanism by which traits are passed on from parents to offspring. Kondrashov's avowal that normal people have within their chromosomes thousands of harmful (deleterious) genes that can be passed on to their children is another ominous bump in the road to Darwinian evolution.

After Kondrashov wrote this article, he had a private conversation with Dr. John Sanford, who will be introduced later in this chapter. Dr. Sanford recorded the following about this conversation (emphases are the author's, not Sanford's)—

"Since this paper, Dr. Kondrashov has indicated to me, by way of personal communication, that ***100 was just a lower estimate, and that 300 is his upper estimate. He also indicated to me that he now believes up to 30% of the mutations may be deleterious.*** *This means that from his perspective "U" (deleterious mutations per person per generation) would be 30-90.* ***This is 100-fold higher than would have previously been considered possible.****"*

Why did Muller, Kondrashov, and the others identified in Kondrashov's quotations above not come to the logical conclusion that, since the human genome is actually degenerating rather than advancing, the Primary Axiom cannot be true?

The answer is because of their religious allegiance to the Axiom. In order to remain loyal to the Primary Axiom, Kondrashov turned to two of Lewontin's "absurd constructs" known as *synergistic epistasis* and *truncation selection* to explain away the very concern that he and Muller had surfaced. Synergistic epistasis and truncation selection will be discussed later in this chapter.

Crow's Concerns

James Crow was Professor Emeritus of Genetics at the University of Wisconsin. He coauthored papers with Motoo Kimura. Sadly, he passed away in January of 2012. Crow specialized in genetic mutations, developed the concept of genetic load, and contributed to the theory of random genetic drift in small populations. His paper[20] on "Genetic Effects of Radiation" published in the *Bulletin of Atomic Scientists* showed that he understood the harmful nature of mutations—

"Even if we didn't have a great deal of data on this point, we could still be quite sure on theoretical grounds that mutations would usually

be detrimental. For a mutation is a random change of a highly organized, reasonably smoothly functioning living body. A random change in the highly integrated system of chemical processes which constitute life is almost certain to impair it – just as a random interchange of connections in a television set is not likely to improve the picture."

In a more recent paper, Crow acknowledged[21] the fundamental problems that high mutation rates cause for the theory of forward evolution (emphases are the author's, not Crow's)—

*"... **The overall impact of the mutation process must be deleterious**... the typical mutation is very mild. It usually has no overt effect, but **shows up as a small decrease in viability or fertility**...**each mutation leads ultimately to one 'genetic death'** ... would surely be an excessive load for the human population... **so we have a problem.**"*

*"... there is a way out ... by judicious choosing, several mutations may be picked off in the same victim ... all individuals with more than a certain number of mutations are eliminated ... **of course ... natural selection does not line up individuals and remove all those with more than a certain number of mutations ...**"*

*"**It seems clear that for the past few centuries harmful mutations have been accumulating ... The decrease in viability from mutation accumulation is 1-2% per generation** ... if war or famine force our descendants to a stone-age life **they will have to be content with all the problems their stone-age ancestors had, plus mutations that have accumulated in the meantime** ... **I do regard mutation accumulation as a problem. It is something like the population bomb, but with a much longer fuse.**"*

Notice how Crow concedes that human beings are now genetically ***inferior*** to our "stone-age" ancestors! Coming from a foreign member of the Royal Society and the former president of both the Genetics Society of America and the American Society of

Human Genetics, this is an astonishing confession about degeneration of the human genome!

Lynch's (et al) Mutation Meltdown

Michael Lynch is Distinguished Professor of Evolution, Population Genetics, and Genomics at Indiana University. Along with Bruce Walsh, he wrote a textbook that is considered the definitive manual of quantitative genetics. Genomic architecture is Lynch's specialty. A paper[22] he coauthored regarding mutation accumulation and the extinction of small populations was published in *The American Naturalist* in 1995. Here are some relevant statements that were made in this interesting paper (emphases are the author's, not Lynch's)—

"As deleterious mutations accumulate by fixation, there is a gradual decline in the mean viability of individuals ... net reproductive rate is less ... ***precipitates a synergistic interaction between random genetic drift and mutation accumulation, which we refer to as mutational meltdown ... the length of the meltdown phase is generally quite short.****"*

"... ***our results provide no evidence for the existence of a threshold population size beyond which a population is completely invulnerable to a mutation meltdown*** *..."*

"... ***simultaneously segregating mutations interfere with each other's elimination by natural selection*** *..."*

The last quotation above is an acknowledgement of *selection interference,* which will be covered in more detail later.

In this paper, Lynch and his coauthors clearly convey that genetic degeneration is a major risk for all currently endangered species. They do not point out, however, that the same would also have applied to many of the species extinctions that have occurred in the past. Nor do they state that extinction from mutation meltdown should also apply to higher genomes. Nevertheless, logic cries out that mutation meltdown applies to both of these situations as well, as shall be covered next.

Higgins and Lynch – More Mutation Meltdown

A paper[23] authored by Michael Lynch and Kevin Higgins and edited by James Crow was published in the *Proceedings of the National Academy of Sciences* in 2001 that is titled "Metapopulation Extinction Caused by Mutation Accumulation." This paper goes far beyond Lynch's 1995 paper in the reach of mutation meltdown (emphases are the author's, not Higgins' and Lynch's)—

"Here we show that metapopulation structure, habitat loss or fragmentation, and environmental stochasticity ***can be expected to greatly accelerate the accumulation of mildly deleterious mutations, lowering the genetic effective size to such a degree that even large metapopulations may be at risk of extinction."*** (Author's notes: a metapopulation is a group of geographically separated populations of the same species that interact, such as the human race. Stochasticity essentially means randomness.)

"... mildly deleterious mutations may create considerably larger mutational load ... because ***individually they are nearly invisible to natural selection, although causing appreciable cumulative reduction in population viability."***

"We find that the accumulation of new mildly deleterious mutations fundamentally alters the scaling of extinction time ... causing the extinction of populations that would be deemed safe on the basis of demography alone."

"Under synchronous environmental fluctuations, the acceleration of extinction caused by mutation accumulation is striking ... ***without mutation, extinction is 2,000 generations ... with mutation accumulation the extinction time is just slightly longer than 100 generations ..."***

"For a metapopulation in unfragmented habitat, ***mildly deleterious mutations are more damaging than highly deleterious mutations*** *... just as in the case with large carrying capacity,* ***the mild mutational effects are the most damaging, causing minimal time to extinction.****"*

"Early work suggested ... accumulation of deleterious mutations may threaten small isolated populations ... here we show that accumulation of deleterious mutations may also be a significant threat to large metapopulations ... ***the decline is sudden, but extinction itself still takes a while to occur, the metapopulation may be completely inviable on the intermediate or long time scales, although appearing healthy on short time scales.****"*

Rather than supporting the theory of forward evolution, this paper makes a case for *backward* evolution via genomic degeneration caused by mutations!

Hoyle's Big Picture

Sir Frederick Hoyle was a British mathematician, physicist, and astronomer. In scientific circles, he is best known for his contribution to the theory of stellar nucleosynthesis. This theory explains how chemical elements heavier than helium are created by stars ejecting much of their mass late in their stellar lifetime. Hoyle is also known for introducing the phrase "Big Bang," although he personally rejected this theory.

Sir Frederick Hoyle was an atheist who believed in Intelligent Design, as explained[24] in this statement—

"If one proceeds directly and straightforwardly in this matter, without being deflected by a fear of incurring the wrath of scientific opinion, one arrives at the conclusion that biomaterials with their amazing measure of order must be the outcome of intelligent design."

And[25] further—

"The notion that not only the biopolymer but the operating program of a living cell could be arrived at by chance in a primordial organic soup here on the Earth is evidently nonsense of a high order."

Hoyle made some other relevant comments about evolution in a book[26] titled *Mathematics of Evolution* that was published in 1999—

"The aging process shows, indeed, that ***statements one frequently hears, to the effect that the Darwinian theory is as obvious as the Earth going round the Sun, are either expressions of almost incredible naiveté or they are deceptions ... with such widespread evidence of senescence*** [aging] ***in the world around us, it still seems amazing that so many people think it 'obvious' that the biological system as a whole should be headed in the opposite direction*** *..."*

"The best natural selection can do, subject to a specific environment, is hold the deleterious mutations in check. ***When the environment is not fixed there is slow genetic erosion, however, which natural selection cannot prevent.****"*

"... natural selection cannot turn back deleterious mutations if they are small, and over a long time a large number of small disadvantages escalate to a serious handicap. ***This long term inability of natural selection to preserve the integrity of genetic material sets a limit to its useful life ...****"*

Unlike the other scientists quoted above, Sir Frederick Hoyle was willing to admit that, based on the findings of his own work, natural selection is *not* omnipotent, and that *"the biological system as a whole"* is headed in a direction opposite from forward evolution!

The author has identified above ten real and significant challenges to the theory of forward evolution that have been raised by seventeen different scientists—all of which are, or were, faithful believers in the Primary Axiom. Remarkably, none of these scientists, except Sir Frederick Hoyle, allowed the results of his scientific work to shake his faith in the Primary Axiom. Rather, each continued to support the Axiom *in spite of* his work. Such is the faith and zeal of modern scientists who have embraced the religion of Darwinian evolution.

For each of the challenges highlighted above, responses have been developed by the evolution community to appease its members. As shall be discussed below, these responses may be placed in Richard Lewontin's category of "constructs of patent absurdity."

In certain situations, the main argument that mainstream contemporary science uses to defend its faith in the Primary Axiom is the number of scientists who have embraced the Axiom. Consider the following from an article[27] in Wikipedia—

"An overwhelming majority of the scientific community accepts evolution as the dominant scientific theory of biological diversity. Nearly every scientific society, representing hundreds of thousands of scientists, has issued statements rejecting intelligent design and a petition supporting the teaching of evolutionary biology was endorsed by 72 US Nobel Prize winners. Additionally, US courts have ruled in favor of teaching evolution in science classrooms, and against teaching creationism, in numerous cases such as Edwards v. Aguillard, Hendren v. Campbell, McLean v. Arkansas and Kitzmiller v. Dover Area School District."

Such an argument reminds the author of the First Council of Nicaea in AD 325 that established the Creed of Nicaea (or Nicene Creed) by a majority vote of the bishops who attended. A significant number of bishops did not vote for this creed, which ultimately resulted in the doctrine of the Trinity.

Rather than objectively evaluating and embracing the consequences of the evolution challenges raised above, the community of Primary Axiom believers has chosen instead to rely on the "votes" of its members to prop up the religion that Darwin founded.

Fear of offending other scientists is also a powerful force for not rejecting the Primary Axiom. The reality of this fear is evident from statements such as—*"If one proceeds directly and straightforwardly in this matter, without being deflected by a fear of incurring the wrath of scientific opinion ..."* (Sir Frederick Hoyle[23]) and—*"It is not antagonistic to the cherished view that evolution of form and function is guided by Darwinian selection ..."* (Kimura[2]).

Obviously, it takes a very brave scientist to honestly, critically, and objectively question the Primary Axiom in the light of the challenges presented above. Such a scientist is Dr. John Christopher Sanford.

Dr. Sanford is a highly successful plant geneticist. Although more than 32 patents have been issued to him, he is probably best known for three inventions—(*i*) the Biolistic Particle Delivery System, or "gene gun," (*ii*) the Pathogen-derived Resistance Process, and (*iii*) the genetic immunization (or vaccination) process. Most of the transgenic crops grown in the world today were genetically engineered with the gene gun technology Dr. Sanford and his collaborators developed. Sanford also started two biotech businesses that are successful today because of his research. He has published more than 70 articles in peer reviewed journals, held an Honorary Adjunct Associate Professor of Botany at Duke University, and is a courtesy associate professor at Cornell, where he previously taught Horticultural Sciences for eighteen years. Dr. Sanford also received the "Distinguished Inventor Award" from the Central New York Patent Law Association in both 1990 and 1995. In conjunction with four other knowledgeable scientists (J. Baumgardner, W. Brewer, P. Gibson, W. ReMine), Dr. Sanford developed a sophisticated open source software package called "Mendel's Accountant" that simulates genetic changes over time.

Like most geneticists, John Sanford became an atheist in his youth. What happened to him later in his life is best told[28] with his own words—

"Late in my career, I did something which for a Cornell professor would seem unthinkable. I began to question the Primary Axiom. I did this with great fear and trepidation. By doing this, I knew I would be at odds with the most 'sacred cow' within modern academia. Among other things, it might even result in my ***expulsion*** *from the academic world. Although I had achieved considerable success and notoriety within my own particular specialty (applied genetics), it would mean I would have to be stepping out of the safety of my own little niche. I would have to begin to explore some very big things, including aspects of theoretical genetics which I had always accepted by faith alone. I felt compelled to do all this—but I must confess that I fully expected to simply hit a brick wall. To my own amazement, I gradually realized that the seemingly 'great and unassailable fortress' which has been built up around the Primary Axiom is really a house of cards. The Primary Axiom is actually an extremely vulnerable theory—in fact it is essentially indefensible. Its apparent invincibility derives largely from bluster, smoke, and mirrors. A large part of what keeps the Axiom standing is an almost mystical faith, which the* ***true-believers*** *have in the omnipotence of natural selection. Furthermore, I began to see that this deep-seated faith in natural selection is typically coupled with a degree of ideological commitment—which can only be described as religious. I started to realize (again with trepidation), that I might be offending a lot of people's religion!*

"To question the Primary Axiom required me to re-examine virtually everything I thought I knew about genetics. This was the most difficult intellectual endeavor of my life. Deeply entrenched thought patterns only change very slowly (and I must add—painfully). What I eventually experienced was a complete over-throw of my previous

understandings. Several years of personal struggle resulted in a new understanding, and a very strong conviction that the Primary Axiom was most definitely wrong. More importantly, I became convinced that the Axiom could be ***shown*** *to be wrong to any reasonable and open-minded individual. This realization was exhilarating, but again—frightening. I realized that I had a moral obligation to openly challenge this most sacred of cows. In doing this, I realized I would earn for myself the intense disdain of most of my colleagues within academia—not to mention very intense opposition and anger from other high places.*

"What should I do? It has become my conviction that the Primary Axiom is insidious on the highest level—having catastrophic impact on countless human lives. Furthermore, every form of objective analysis I have performed has convinced me that the Axiom is clearly false. So now, regardless of the consequences, I have to say it out loud: ***the Emperor has no clothes!***

"If the Primary Axiom is wrong, then there is a surprising and very practical consequence. When subjected only to natural forces, the human genome must irrevocably degenerate over time. Such a sober realization should have more than just intellectual or historical significance. It should rightfully cause us to personally reconsider where we should rationally be placing our hope for the future."

True to his moral obligation, Dr. Sanford wrote a book titled *Genetic Entropy & The Mystery of the Genome* that was published in 2005. An updated version was released in 2008. These books contain new science that exposes the truth that the Primary Axiom is wrong.

After the author read Dr. Sanford's book, he realized that he had finally found the real scientific confirmation of his faith in God as The Creator that he had been seeking. Consequently, the author recommends *Genetic Entropy & The Mystery of the Genome* to all Christians—whether scientists or not.

The remainder of this chapter summarizes the results of Dr. Sanford's work that are detailed in his enlightening book. Much of what the author has written below comes from the book,[29] with Dr. Sanford's approval.

The Human Genome

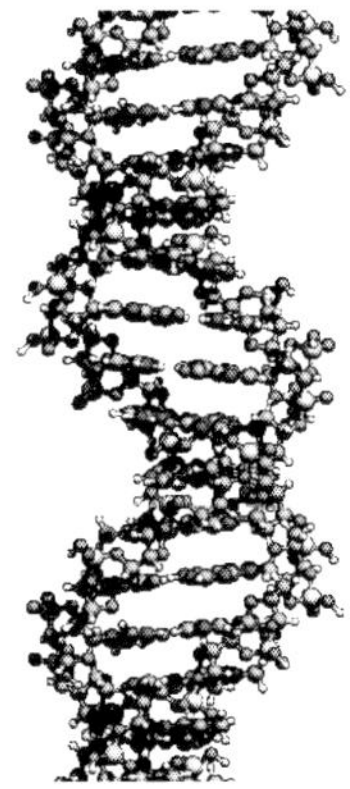

DNA
Structure
courtesy Richard Wheeler

According to Dr. Sanford, the human genome is the instruction manual that specifies human life. It instructs cells to be cells and the human body to be the human body. There is no information system designed by man that can even begin to compare with it. Science has so far only discovered the first dimension of this "book of life," which is a linear sequence of molecules called *nucleotides*. These tiny molecules make up the individual steps of the spiral staircase structure of DNA (shown here), and they form the letters of the genetic code. Small clusters of these four molecular letters make up the words of the instruction manual, which combine to form *genes* (the chapters of the manual), which in turn combine to form *chromosomes* (the volumes of the manual), which also combine to form the entire *genome* (the entire library that contains all the volumes).

A complete human genome contains dual sets of *3 billion* nucleotides each. These linear codes overlap to form an extremely sophisticated information system that even utilizes data compression. Besides employing multiple overlapping, linear language-like forms of genetic information, the genome also contains innumerable loops and branches—similar to a complex computer program. A hierarchy of genes provides regulation of genes on multiple levels. Certain genes can sense changes in the environment and then instruct other genes to react by setting in motion complex cascades of events that can then modify the environment. Some genes actively rearrange themselves, or modify other gene sequences—essentially *changing* portions of the instruction manual! Of course, this is how adaptive evolution proceeds.

There is solid evidence that linear DNA can actually fold into two- and three-dimensional structures, which likely encode even higher levels of genetic information. Within a typical non-dividing nucleus of a human cell, there is reason to believe that there exists incomprehensibly complex three-dimensional arrays of DNA, whose 3D architecture further controls higher biological functions.

Thus the human genome's set of instructions is dynamic, self-regulating, and multi-dimensional. The genome's higher levels of complexity and interaction are probably beyond the reach of mortal understanding, given the incredible complexity of life.

All this genetic information is packaged within a cell's nucleus, which is so tiny it can only be seen with a powerful microscope. The human body contains more than *100 trillion cells*, and each cell contains a complete set of genetic instructions that include its own highly prescribed duties.

The human genome not only specifies the complexity of cells and the body, it also specifies the functioning of the brain—a still higher level of organization that is entirely beyond mortal comprehension.

From where did all this fabulously complex information come?

According to Darwinian evolution, *mutation* and *natural selection* have created all the biological information on the earth today. All plant and animal genomes (or instruction manuals) must have evolved from the simple "first" genome by way of a long series of mutations (or typographical errors) and lots of natural selection (or differential copying). This is the Primary Axiom of biological evolution. Stated another way, *life is life because random mutations at the molecular level are filtered through a reproductive sieve on the level of the whole organism.*

It is important to understand that genetic mutations occur at the molecular level, but natural selection can only occur at the level of a complete organism. This truth has important ramifications for the validity of the Primary Axiom. Furthermore, mutations occur on a random basis, without any intelligent interaction, and selection translates to the mere tendency for some organisms to reproduce more than others, again without intelligent guidance.

There are two major causes of nucleotide mutation—errors which occur when a cell copies its DNA in preparation for cell division, and damage that results from environmental agents such as nuclear radiation, ultraviolet light in sunshine, and certain chemicals.

Distribution of Random Mutations

Genetic mutation is catastrophic.

Considering all human genetic predispositions to disease, it is clear that all members of the human family are highly mutant. Nearly every family is, or has been, impacted in some way by the tragedy of cancer, which is fundamentally the end result of body cell mutations. Furthermore, there is growing evidence that aging itself is caused by the accumulation of body cell mutations. Globally, health improvement campaigns and policies attempt to reduce or minimize mutation, thereby reducing the risk of cancer and other degenerative diseases.

How can mutations be considered good?

The Primary Axiom implies that mutations are good because they provide the variation in organisms that enables natural selection, and thus forward evolution, to take place—ultimately creating the genetic (biological) information required for life.

Furthermore, the Primary Axiom requires that mutations be random, since no genetic variation by design is allowed.

For purposes of assessing the accuracy of the Primary Axiom, a mutation should only be considered beneficial if it *creates new genetic (biological) information*. While many mutations have been declared beneficial by science, most of them actually destroyed biological information rather than create it. As of the 2005 printing of his book, Dr. Sanford had not seen a single, crystal-clear example of a known mutation that unmistakably created genetic information.

There are more than 3 billion potential mutation sites within the human genome. Only a small fraction of these nucleotide positions, if mutated, will have a major effect on human fitness one way or the other. However, no potential mutations in any of the sites can be conclusively shown to have no effect. Therefore, the vast majority of nucleotide positions are considered *nearly-neutral* sites, meaning that a mutation in one of these locations will be only slightly *deleterious* (causing harm or damage). As was discussed earlier in this chapter, Motoo Kimura published "The Neutral Theory of Molecular Evolution" in 1968 to establish the reality of this concept. Today, virtually all population geneticists agree that most mutations are nearly-neutral.

On the other hand, no nucleotide position can be considered to be completely neutral, since it consumes space and affects the spacing between other sites. Even if a nucleotide were to carry absolutely zero genetic information (probably not possible), it would still be slightly deleterious because it would slow cell replication and waste energy.

Thus, the reality of random mutations is that they are almost exclusively slightly deleterious. Beneficial mutations are so rare they are typically never even shown in mutation distribution graphs.

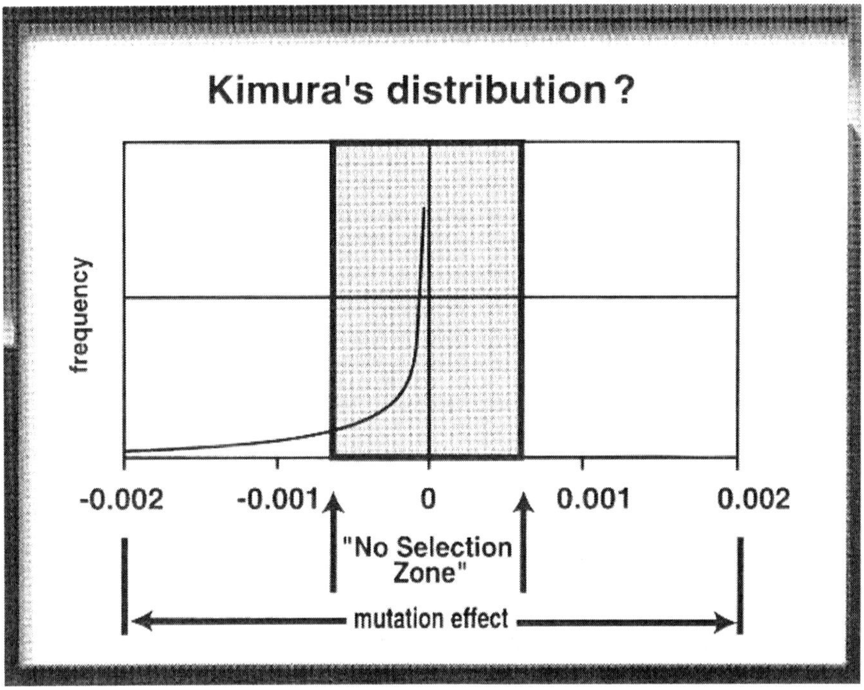

Figure 1

The graph shown in Figure 1 is based on Kimura's work[31] published in 1979. Dr. Sanford modified and expanded Kimura's curve to simplify its message for readers who are not trained geneticists. The curve in Figure 1 very nearly represents the true distribution of random mutations.

Where the mutation effect value is negative, the mutation is deleterious (harmful, not beneficial). Notice that most of the deleterious mutations are placed near the zero mutation effect, illustrating Kimura's assertion that they are nearly-neutral. However, none of them actually have a value of zero, as discussed above. Kimura refers to nearly-neutral mutations as being "effectively neutral," meaning that they are so subtle they are not subject to natural selection. He became famous for introducing the shaded box in the graph, which can be thought of as the "no selection zone." Remember that this graph represents the work of a population geneticist who believed in the Primary Axiom.

Kimura does not show any beneficial mutations (i.e., to the right of zero in the Figure 1 graph) in his own published graph. Evidently he considered beneficial mutations to be so rare they were not worthy of his consideration in the 1979 paper . . . or perhaps he had another reason, as discussed below.

Given Kimura's distribution of mutations shown above, one wonders how scientists can possibly explain and believe in forward evolution. They do it by redefining everything in the no selection zone to be *completely neutral* rather than *nearly-neutral*. By doing so, they summarily dismiss and ignore all these slightly deleterious mutations! As discussed above, no mutations are completely neutral. Even Kimura is careful in his 1979 paper to show that no mutations touch the zero line. And yet, evolution scientists conveniently ignore all these mutations, which will be shown later to be of critical importance. The author considers this unscientific approach to belong in Richard Lewontin's category of "constructs of patent absurdity."

With all slightly deleterious mutations in the no selection zone ignored, evolution scientists assume that all mutations to the left of the no selection zone (called *minor* mutations) can be entirely eliminated with natural selection. Thus, with all deleterious mutations so eliminated, these scientists argue that, no matter how rare beneficial mutations that are to the right of the no selection zone might be, there should be enough time and selection power available to use these beneficial mutations as the catalysts of forward evolution. It will be shown later that these scientists are wrong on both accounts.

Even though Kimura did not show any beneficial mutations in his 1979 paper, Dr. Sanford felt it would be enlightening to see how the beneficial mutation curve would compare with Kimura's deleterious mutation curve. Referring to the previously discussed 1997 paper of James Crow, Dr. Sanford confirmed that the beneficial mutation curve should essentially be a mirror image of the deleterious mutation curve in terms of shape (but not value). He also knew the beneficial mutation curve would have a dramatically lower frequency of occurrence. A 1998 paper[32] by Phillip Gerrish and Richard Lenski estimates the frequency ratio of deleterious to beneficial mutations to be *one million to one*. Furthermore, the actual frequency of beneficial mutations occurring is so infinitesimally small it defies actual measurement. In a different paper[33] published in 2000, Thomas Bataillon said this about beneficial mutations—

"Yet most experiments looking at multicellular organisms have so far failed to produce any information on such mutations."

Using such information published by evolution scientists, Dr. Sanford enhanced Figure 1 to include a curve for beneficial mutations, as shown in Figure 2 on the next page. With a deleterious to beneficial mutation ratio of a million to one, the size of the beneficial mutation curve shown in Figure 2 is much too large; however, it is as small as Dr. Sanford could make it while keeping it visible.

The most important revelation that can be gleaned from this graph is that **virtually all beneficial mutations fall within the no selection zone! This means that natural selection cannot favor beneficial mutations!**

Without such selection, beneficial mutations will essentially drift out of the genome of the human race. Dr. Sanford was shocked when he came to this realization. One may speculate that this is the *real* reason Kimura did not include the distribution of beneficial mutations in his 1979 paper.

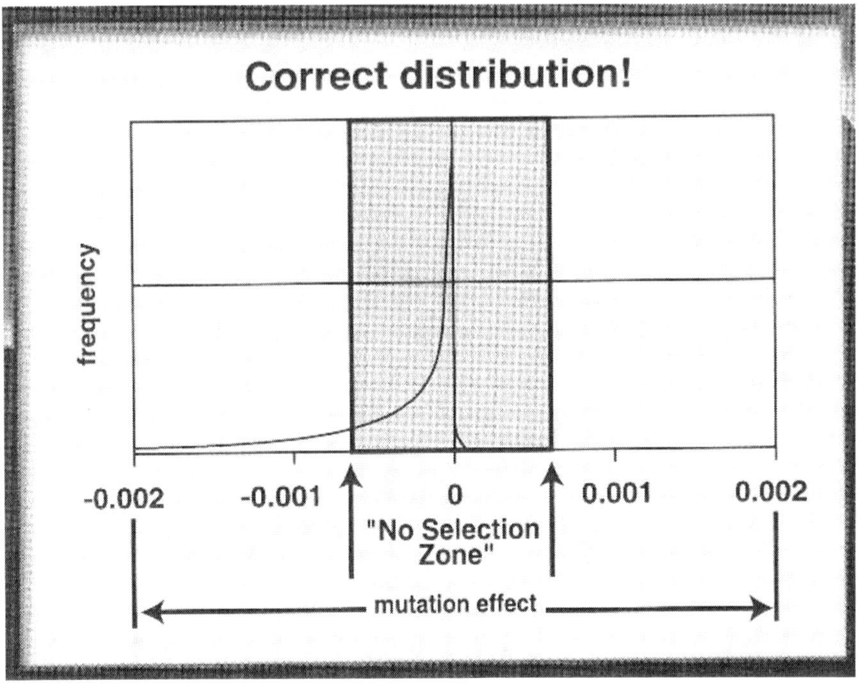

Figure 2

Dr. Sanford's enhanced graph illustrates why mutations cannot result in a net gain of genetic (biological) information. The overwhelming preponderance of deleterious mutations in the no selection zone guarantees a net loss of information rather than a gain.

Everything about the true distribution of mutations argues against their possible role in forward evolution.

Furthermore, a massive mutation study confirmed this, as explained by Dr. Sanford—

"During the last century, there was a great deal of effort invested in trying to use mutation to generate useful variation. This was especially true in my own area—plant breeding. When it was

discovered that certain forms of radiation and certain chemicals were powerful mutagenic agents, millions and millions of plants were mutagenized and screened for possible improvements. Assuming the Primary Axiom, it would seem obvious that this would result in rapid 'evolution' of our crops. For several decades this was the main thrust of crop improvement research. Vast numbers of mutants were produced and screened, collectively representing many billions of mutation events. A huge number of small, sterile, sick, deformed, aberrant plants were produced. However, from all this effort, almost no meaningful crop improvement resulted. The effort was for the most part an enormous failure, and was almost entirely abandoned. Why did this huge mutation/selection experiment fail—even with a host of Ph.D. scientists trying to help it along? It was because even with all those billions of mutations, there were no significant new beneficial mutations arising.

"If essentially no truly beneficial mutations (i.e. resulting in more net information) could be recovered from this vast science-guided process, why do we think the identical process, in the absence of any guiding intelligence, would be more fruitful in nature? However, the very same scientists who failed at mutation/selection were ***extremely*** *successful in crop improvement when they abandoned mutation breeding and instead used the pre-existing natural variation within each plant species or genus. This would make sense if such pre-existing variation did not principally arise via mutation—but originally arose by* ***design****."*

Mutation Accumulation

As discussed previously in the sections Muller's Fear and Crow's Concerns, geneticists have been worried about the impact of mutation on the human race for many decades. When the concerns first surfaced, the rate of deleterious mutation was estimated to be from 0.12 to 0.30 mutations per person per generation. The belief then was that, if the rate approached one deleterious mutation per person per generation, long-term genetic deterioration would be a certainty.

In 2000, Michael Nachman and Susan Crowell estimated the rate to be 175 deleterious mutations per person per generation.[34] As explained in the Kondrashov's Numbers section above, Dr. Kondrashov told Dr. Sanford personally in 2002 that the rate may be as high as 300 deleterious mutations per person per generation, and that 100 was his low estimate.

All these numbers are too low because they only consider point mutations—which are mutations that occur in a single nucleotide position. Other types of common mutations include deletions, insertions, duplications, translocations, and inversions—all of which involve multiple nucleotide positions. There are also other types of non-point mutations that are not covered here. Without going into the details that Dr. Sanford provides in his book, a more accurate estimate of the mutation rate is **more than 1000 nucleotide changes in every person every generation!**

How much of the human genome is functional, and therefore susceptible to this amazingly high mutation rate? Discussing the answer to this question is beyond the scope of this book, but Dr. Sanford's book covers it in detail. Dr. Sanford summarizes the answer this way—

"It is becoming increasingly clear that most, or all, of the genome is functional. Therefore, most, or all, mutations in the genome must be deleterious."

What does this mean? Assuming a very conservative mutation rate of only 100 deleterious mutations per person per generation, Dr. Sanford provides the answer—

"On a person basis, 100 mutations represent a loss of only a miniscule fraction of the total information in our genome (the genome is

huge). However, the real impact of such a high mutation rate will be at the population level, and is primarily expressed with the passage time. Since there are six billion people in the world, and each person has added an average of 100 new mutations to the global population, our generation alone has added roughly 600 billion new mutations to the human race. If we remember that there are only three billion nucleotide positions in the human genome, we see that in our lifetime there have been about 200 mutations for every nucleotide position within the genome. Therefore, every possible point mutation that ***could*** *happen to the human genome* ***has*** *happened many times over—just during our lifetime! Because of our present large population size, humanity is now being flooded by mutations like never before in history. The consequences of most of these mutations are not felt immediately, but will become manifested in coming generations."*

Natural Selection Limitations

Most human geneticists agree that the human race is currently degenerating genetically due to rapid deleterious mutation accumulation and relaxed (i.e., reduced) natural selection. They realize that such degeneration reduces the overall fitness of the race. The decline in fitness is currently believed to be occurring at a rate of 1-2% per generation.

So why is average lifespan currently increasing rather than decreasing?

The answer is that the degeneration is happening at the genetic level while medical and technical advances are acting at the individual person level to increase average life spans. Genetic degeneration requires multiple generations before manifesting reduced fitness, which appears in its final stages in the form of declining fertility. Meanwhile, medical technology has been effective in preventing or delaying death from the five leading causes—heart disease that reduces blood supply, stroke, lower respiratory infections, chronic obstructive lung disease, and diarrheal diseases. Treatments as simple as diet change have extended the lifetimes of many people around the world. However, medical treatments like regular exercise, diet change, antibiotics, vaccinations, medications that lower hypertension, etc. have no effect on genetic degeneration. Over time, degeneration of the human genome will take its toll on the human race. Each new generation will add more deleterious mutations to the genome, accelerating its downward slide.

The reader may ask why natural selection will not halt the downward slide. The answer stems from the fact that genetic mutations occur at the molecular level while natural selection can only occur at the level of human reproduction—i.e., at the level of individual human beings. This is an enormous chasm.

In order to stop the downward slide, selection must occur for billions of extremely subtle and complex genetic differences at the molecular level. Natural selection cannot see the molecular level of a person; it can only see complete human beings. So how will natural selection be able to do the job?

To illustrate the problem, imagine yourself buying a new television. You want to get the very best one that is available in the store. To do this, you should visually examine each pixel on each television screen to search for defects. In parallel, you should examine every component on every circuit board to identify electronic defects. And somehow, you should test each component to determine if it functions properly, and then test each circuit board as well. Of course, you will need to correlate your findings of pixel and electronic defects in order to identify the best television in the store. Is there any way you can do all this? Of course not! Instead, you simply watch several different televisions tuned to the same channel for a few moments and then choose the one that appeals to you the most. Pixels and electronics are never even considered when you make your decision.

The biological analogy of this illustration is even more complex. It involves the equivalent of pixels within pixels, within pixels, within pixels; also very long chains of events that effectively separate the impact of a single nucleotide at the molecular level from its distant and highly diminished influence on the entire person. At each step of these chains, there is dilution of the nucleotide level impact, resulting in an order of magnitude loss of resolution and relevance. Natural selection might as well be trying to measure the force of a butterfly's wing stroke on the eye of a hurricane that is a thousand miles away!

There is another miracle of life that comes into play here. It is called *homeostasis*—a process which maintains the stability of the human body's internal environment in response to changes in external conditions. As a simple example, a healthy body will attempt to maintain its internal temperature regardless of the temperature of the surrounding environment. If the body senses that the external temperature is becoming uncomfortably warm, it will automatically commence perspiration to cool itself down. In contrast, shivering is the technique it utilizes to warm itself if early hypothermia is detected.

Homeostasis is functional and highly effective in all organisms, and is implemented with a complex network of sensors and regulators *within each cell*. The network operates on every level of the organism—from molecules to the entire body—and makes it even more difficult for natural selection to select for anything below the level of an entire organism.

If the Primary Axiom were true, natural selection would be able to see all six billion nucleotides in a single person and select them (or not) on an individual nucleotide basis. Yet Darwin would be the first to admit that natural selection only selects at the level of *entire* human beings—in essence, whether they reproduce or not.

Darwin did not know about nucleotides or genetics. Gregor Mendel, the Austrian monk who laid the mathematical foundations of genetics, was his contemporary. It was not until 1918—thirty-four years after they had both passed away—that Mendelian genetics and Darwin's *On the Origin of Species by Means of Natural Selection* started to come together. Three scientists—Jack Haldane, Ronald Fisher, and Sewall Wright—developed the science of population genetics from 1918 to 1932. They integrated genetics with natural selection and provided a foundation for *modern evolutionary synthesis,* which was subsequently advanced between 1936 and 1947.

The genetic scientists who moved the theory of evolution forward during this time realized that they had a problem with the chasm between natural selection and molecular mutations. Since they were all religiously committed to Darwinism, they had to devise a way to make forward evolution *appear* genetically feasible. How they did this is best described by Dr. Sanford—

"To overcome the problem, they very cleverly transferred the ***unit of selection*** *from the whole organism to the genetic unit (i.e. the gene or nucleotide). To do this they had to re-define a population of people as being essentially a 'pool of genes.' In this way, they could claim that real selection was operating at the level of each individual nucleotide within the pool—not really on people at all. Each individual nucleotide could be envisioned as being independently selected for, or against (or neither). This made it very easy to envision almost any evolutionary selection scenario, no matter how complex the biological situation. This was an incredibly effective way to obscure the entire problem. Indeed, Darwinism would have died very naturally at this point in time, except for this major intellectual invention."*

Dr. Sanford continues—

"There is one serious problem with re-definition of the problem in this way. That problem is that the new picture is categorically false.

Populations are not even remotely like pools of genes, and selection is ***never, ever*** *for individual nucleotides. To justify this radical new picture of life, the theorists had to axiomatically assume a number of things—which were in fact all known to be clearly false. For example they had to assume that all genetic units could assort independently—so each nucleotide would be inherited independently—as if there were no linkage blocks (totally false). Likewise, they had to assume no epistasis—as if there were no interactions between nucleotides (totally false). They also typically assumed essentially infinite population sizes (obviously false), they usually implicitly assumed unlimited time for selection (obviously false), and they generally assumed the ability to select for unlimited numbers of traits simultaneously (which we will show to be false). So from the very beginning of population genetic theory many unrealistic and unreasonable assumptions needed to be made, to make the model appear even feasible.*

"On this very false foundation was built the theoretical pillars of modern population genetics. The theorists' models did not match biological reality, but these men had an incredible aura of intellectual authority, their arguments were very abstract, and they used highly mathematical formulations which could effectively cow most biologists. Furthermore, most biologists were also committed Darwinists, and so were philosophically in agreement with the population geneticists. So most biologists were more than happy to go along for the ride, even if the story did not quite make sense. In fact, the early population geneticists quickly became the 'darlings of science' and were essentially idolized.

"What is the biological reality—apart from ideology? The reality is that selection acts on the level of the organism, not on the level of the nucleotide. Human genes never exist in 'pools,' they only exist in massive clusters, within real people. Each nucleotide exists intimately associated with all the other nucleotides within a given person, and they are only selected or rejected as a set of 6 billion."

Along with Dr. Sanford, the author considers this unscientific (pseudoscience!) approach for salvaging forward evolution to belong in Richard Lewontin's category of "constructs of patent absurdity."

Since the concept of gene pools was invented, scientific research has shown that the problem it tried to gloss over is actually much worse than the early population geneticists could have even imagined—the size and complexity of the human genome is much more vast, homeostasis is much more extensive, and there are many more levels of organization separating the genetic makeup and the observable traits of a human being than were previously known.

Clearly, the Primary Axiom fails at this fundamental level, and therefore should be rejected by anyone seeking the truth.

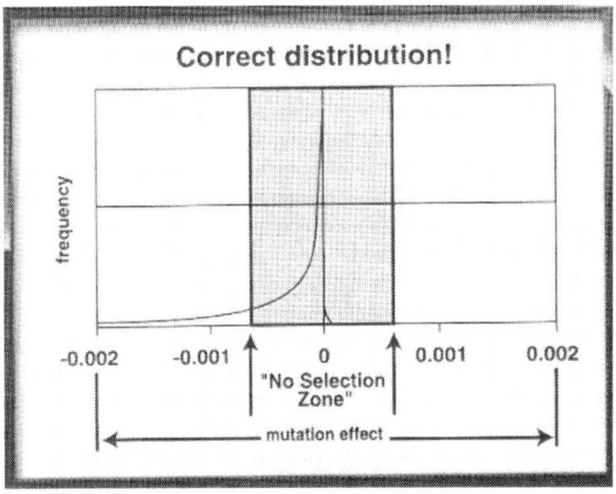

Figure 2

There is, however, a category of mutations for which natural selection works, and works well. It is the category of *minor* mutations, which are located just to the left of the no selection zone in Figure 2. Remember that the majority of mutations, which are nearly-neutral (as opposed to minor), reside within the no selection zone. Minor mutations have a subtle but distinct effect on reproductive capability, and therefore directly contribute to natural selection. In addition, the effects of minor mutations are partly additive, so natural selection can actually select for multiple minor mutations simultaneously. Thus, for minor mutations, natural selection works elegantly and appears to have been designed to stabilize life, which otherwise would deteriorate very quickly. In this sense, natural selection is a marvelously designed system!

However, natural selection for minor mutations has significant limitations. Foremost is the concept of *biological noise*, which can disrupt the otherwise natural elimination of minor mutations. The concept of biological noise is best understood by comparing it to the noise, or static, that one often hears when listening to an AM radio. Hearing such static when trying to enjoy music or get the latest news is very annoying because it results in the loss of information (whether musical or verbal). Radio noise is an interfering signal that diminishes the clarity of a radio broadcast. It can come from a variety of sources, including other radio stations, solar flares, cosmic radiation, cell phones, electric motors, power lines, etc. Regardless of the source, radio noise always causes loss of information. Therefore, in order to minimize the amount of information lost, the ratio of radio broadcast signal to radio noise (i.e., the signal-to-noise ratio) must be maximized.

So it is with biological noise, where signal-to-noise ratio is measured in terms of *heritability* (how likely a trait is to be passed on from one generation to another). If a trait has high heritability, it will be easy for natural selection to select it. If it has low heritability, the opposite is true. Most nucleotides are not selectable because of consistently low signal-to-noise ratios. Furthermore, attempting to select many nucleotides simultaneously always fails because of even lower net signal-to-noise ratios that result. A primary reason why natural selection cannot stop genome deterioration is ubiquitous biological noise. Attempting to apply natural selection to the entire human genome causes the signal-to-noise ratio to approach zero. Biological noise consistently outweighs the effects of individual nucleotides. This is a fundamental reason why natural selection may work on the level of a gene, but fails on the level of the genome.

While the effects of minor mutations on reproductive capability are very subtle, the effects of biological noise factors can be very prominent. For example, widespread deaths caused by a war will clearly overshadow the vastly subtler effects of minor mutations when it comes to influencing human reproduction. Furthermore, the fact that most mutations are *recessive* (they require the same mutant gene to come from both parents in order to be expressed) severely masks their negative fitness effects and greatly hinders natural selection against them. Also, the interaction between genes and their modifier genes (*epistasis*) interferes with selective elimination of minor mutations. All these effects cause the occurrence of *genetic drift*, which means that gene frequency changes in a population (usually downward) are determined (statistically) by random occurrences rather than by natural selection.

When Motoo Kimura defined his "no selection zone" box (see Figure 1), he sized it considering only one type of biological noise—gamete sampling. Not surprisingly, he did not take into account the *primary* sources of noise. When they are included, the size of the near-neutral mutation box must be enlarged, as shown in Figure 3 on the next page. The first expansion accounts for the poor heritability of mutations, which results from the fact that most variation in a population is caused by physical environment rather than genetic variation. The second expansion accounts for the poor correlation between trait superiority and reproductive success (i.e., human reproduction tends to be more random than based on partner fitness).

When both of these major sources of biological noise are factored in to resize the no selection zone, the box virtually doubles in width—and thus the percentage of mutations (both

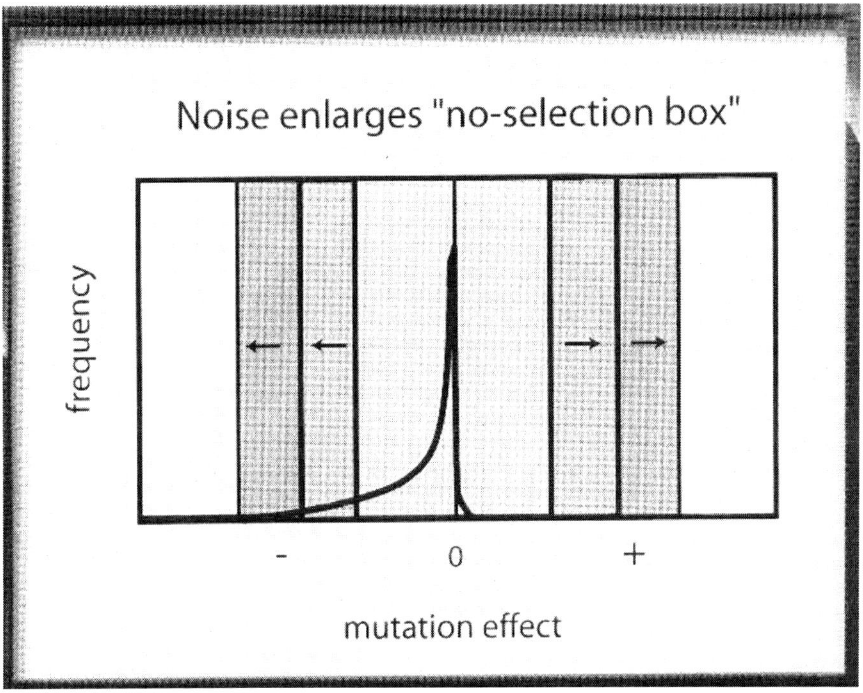

Figure 3

beneficial and deleterious) that natural selection cannot select grows accordingly. A direct consequence of the wider box is that more minor mutations are un-selectable, further limiting the overall effectiveness of natural selection.

While natural selection can work for multiple minor mutations, the fitness effect of each mutation diminishes as the number of mutations increases, thus lessening the significance of that mutation's effect on total reproductive probability. As the number of minor mutations increases, a threshold is reached where each mutation's impact effectively becomes nearly-neutral and natural selection therefore becomes ineffective. This phenomenon means that genomic selection involving millions of minor mutations will always fail.

Another limitation of natural selection is known as *selection interference*. Selection interference occurs when selection for one trait interferes with selection for another trait. Within an individual, a desirable trait will almost always coexist with one or more undesirable traits. Therefore, to select against an undesirable trait automatically means unintentionally selecting against a desirable trait—since selection must accept or reject a whole person. This means that natural selection can never operate on a single mutation in isolation. To select for a beneficial mutation will always result in the selection of many, many more undesirable traits. While selection interference is often alluded to by population geneticists, Dr. Sanford believes it is a problem for forward evolution that has not been adequately addressed.

Clustered Mutations

Virtually all of the human genome exists in *linkage blocks*, which can be thought of as genome "building blocks." A linkage block consists of clusters of nucleotides that are physically linked and inherited as if a single trait. Referring to Figure 2, the number of beneficial mutations in the human genome is vastly outnumbered by deleterious mutations (on the order of a million deleterious mutations per single beneficial mutation). Because of this ratio, any beneficial mutant nucleotide that might reside in a given linkage block will be surrounded by thousands of deleterious mutations. Thus, every beneficial mutation is inseparably tied to a large number of deleterious mutations. Since clusters in a linkage block never break apart, the net effect of a cluster of mutations is that it will be inherited as if a single mutation—having the *net effect* of all its component mutations. Thus beneficial mutations will always be canceled out by a preponderance of physically linked deleterious mutations.

The actual distribution of linked mutation clusters is represented by the shaded area in Figure 4 below. Notice that this shaded area is shifted substantially to the left compared to the distribution curve of single mutations shown with it, which means that the average magnitude of harmful effect is typically greater for clustered mutations than for individual mutations. Also, the frequency of clustered mutations is much higher than that of single mutations where the deleterious effects are most damaging.

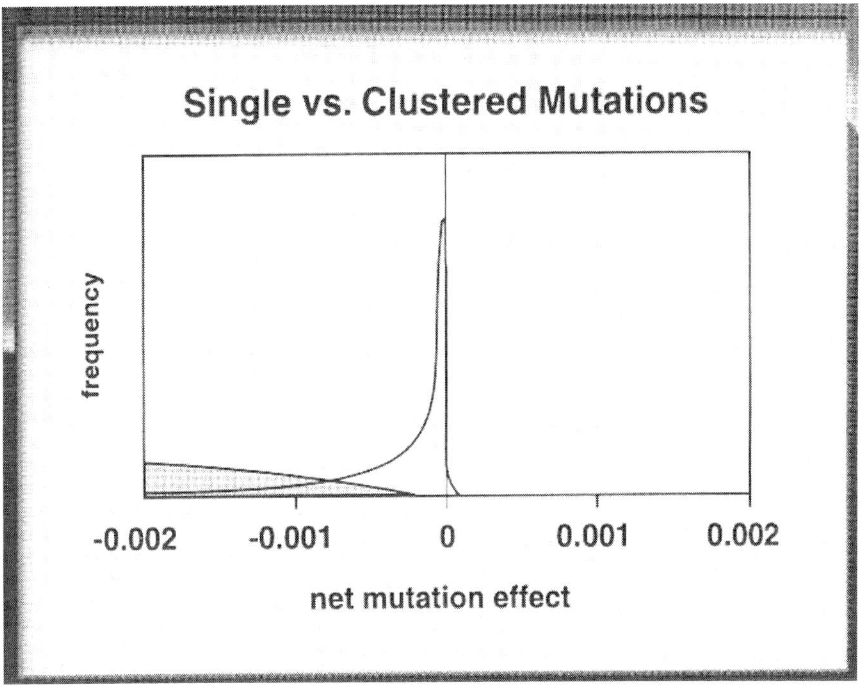

Figure 4

The human genome is a composite of some 100,000 to 200,000 linkage blocks. From the distribution of clustered mutations shown in Figure 4, it is clear that all these "building blocks of evolution" are deteriorating. This is another proof that the human genome is degenerating rather than gaining new genetic information to support forward evolution.

Natural selection is essential, but not in the way Darwinian evolutionists claim; rather, it is vital for *slowing down* the degeneration and for supporting adaptive evolution. Unfortunately, there is no form of natural selection that can actually halt the accumulation of harmful mutations in the human genome.

Evaluating Crow's Solution

The scientific analyses presented so far solidly argue that *net information* in the human genome must be declining each generation. Therefore, the genome could never have arisen spontaneously. This logical conclusion exposes the fallacy of the Primary Axiom and the myth of forward evolution.

James Crow understood what genome deterioration could mean for the Primary Axiom. Therefore, he proposed a natural selection model in 1997 that was designed to salvage the Axiom. This model is explained in one of his quotes from the section above titled Crow's Concerns—

"... there is a way out ... by judicious choosing, several mutations may be picked off in the same victim ... all individuals with more than a certain number of mutations are eliminated ..."

Crow's model assumes that the randomness of mutations will result in certain people having considerably more mutations than others. His solution is to focus selection only against these highly mutant people, thereby eliminating a disproportionate number of mutations. The model uses the concept of *truncation selection,* which artificially selects individuals strictly according to their mutation count. Crow's claim was that more mutations in the human race would be eliminated at lower "cost" to the race using this approach. He theorized that, eventually, the number of mutations per person might be stabilized and thus the decline in fitness would taper off.

The graph shown in Figure 5 below provides a computer simulation of Crow's model. As seen in this graph, average mutations per person accumulate to high numbers as generations pass, but they do, although slowly, start to taper off.

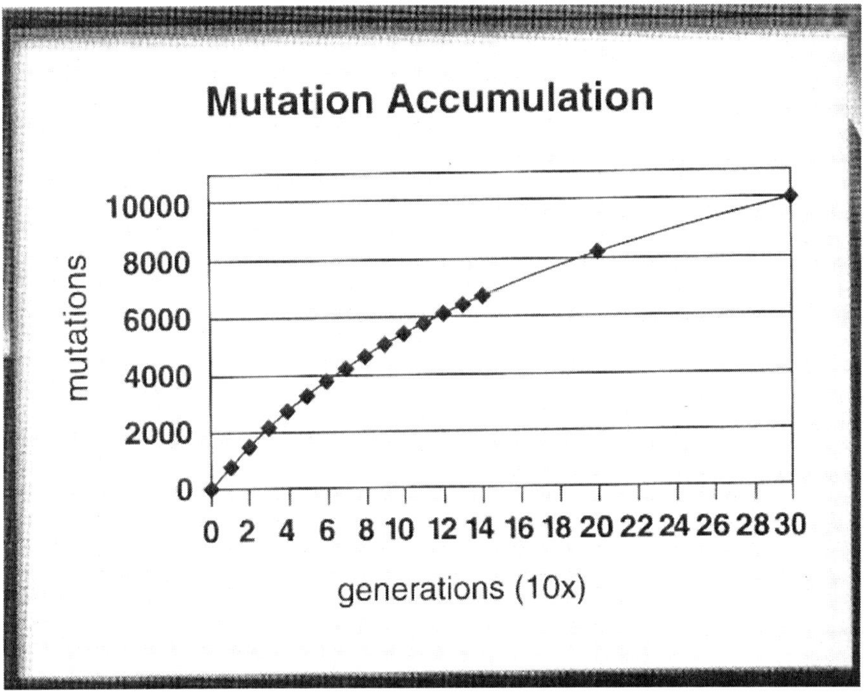

Figure 5

Crow's model was intended to eliminate the problem that mutation accumulation causes for the Primary Axiom. Considering the assumptions that he uses—

- all mutations have equal value
- all mutations are individually very subtle
- no mutations are nearly-neutral
- all selection is based on mutation count
- artificial truncation selection is naturally operative

—*none is even remotely reasonable!* Therefore, Crow's model belongs in Richard Lewontin's category of "constructs of patent absurdity" and is another example of pseudoscience.

Even though Crow used artificial and unreasonable assumptions to accomplish his goal of propping up the Primary Axiom, the graph in Figure 5 still shows serious genetic damage. Using data from this simulation, along with the conservative assumption that the average mutation reduces fitness by one part in 10,000 (i.e., 0.0001), average population fitness is simulated in the graph shown below in Figure 6. Notice that the fictitious species in this graph begins to go extinct in 300 generations (about 6,000 years). **Thus Crow's model, even based on totally unrealistic assumptions, does not support the Primary Axiom after all!**

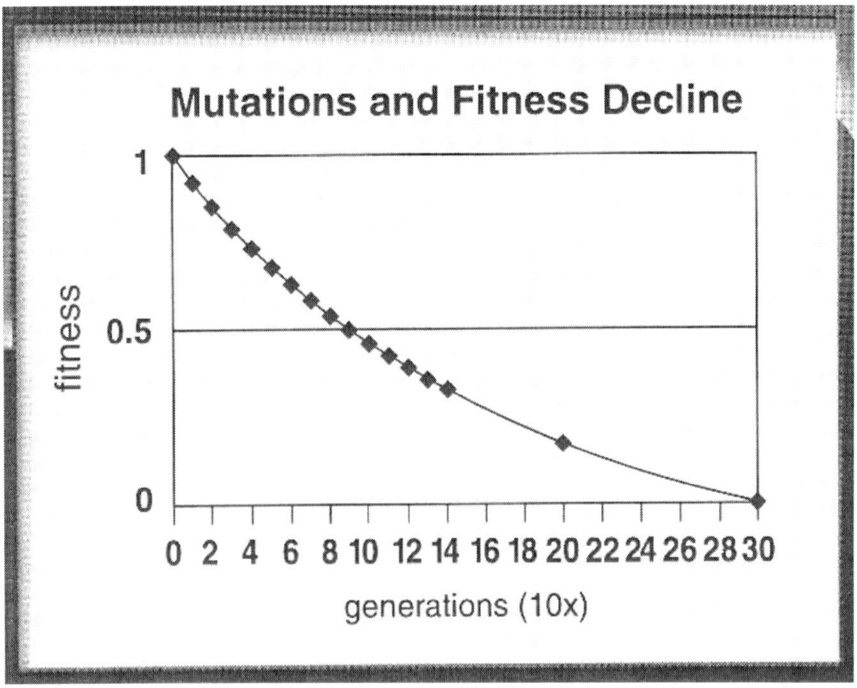

Figure 6

If one assumed that the average mutation reduced fitness by much less than 0.0001, then all the mutations would become effectively neutral, and therefore would become unselectable. For this case, Crow's model breaks down altogether.

Independent work[35] by D. Schoen, J. David, and T. Bataillon that was published in 1998 showed similar results to both of these simulations.

There are two ways to classify who is "most mutant." Crow uses the first approach in his model, which is selecting for those who *numerically* have the most mutations. This method does not reflect the real world since each of us is made up of genes that essentially represent a random sampling of the entire human race's gene pool. Genetic mutations have been accumulating and mixing within this pool for hundreds of generations, so mutation count per person should be fairly homogeneous by now, and individual variations from the average should only be minor. One might wonder why Crow chose this approach for his model. It has the appearance of a mathematical trick designed to obscure or mystify a major theoretical problem.

The second approach is to consider those who have the *worst* mutations. In the human genome, the fitness impact of different nucleotides can vary by multiple orders of magnitude, and can range from nearly-neutral to lethal. A lethal mutation reduces fitness 100%—meaning there can be no reproduction of an individual who has a lethal mutation. At the other extreme of the range, nearly-neutral mutations reduce fitness by only a millionth of a percent or even less. In between, minor mutations degrade fitness by 1% or 2%. This huge disparity in range means that the major differences between human beings are not due to the total number of mutations, but instead are due to the specific effects of a relatively few *high-impact* genetic differences. Just one minor mutation can overshadow the effects of a million nearly-neutral mutations, and one lethal mutation will overshadow the effects of all other mutations—minor and nearly-neutral.

It should be evident that selecting away individuals with the highest counts of mutations is not a reasonable natural mechanism to eliminate mutation accumulation. A limited number of relatively major nucleotide variants are the real basis for the obvious differences between individuals. Therefore, this relatively small number of "significant" nucleotides must be the real foundation for both adaptive evolution and artificial breeding of plants and animals. While selection can progress rapidly for a handful of variants at these major nucleotide sites, the human genome is still degenerating as a whole.

Evolution scientists have coined the sophisticated-sounding phrase *synergistic epistasis* to further mystify this challenge to the Primary Axiom. As mentioned previously, epistasis involves mutated genes interacting with other modifier genes. Therefore, "synergistic" epistasis must mean that mutations interact in such a way that several mutations cause more damage collectively than they would individually. At least one paper[36] written by Santiago Elena and Richard Lenski provides experimental evidence that the concept of synergistic epistasis is not valid. But even if it were, it would actually make the situation worse, not better, for forward evolution! Genetic scientists have always known that genetic units interact, and that epistasis is a major obstacle to natural selection because it creates biological noise. If synergistic epistasis were truly a valid concept, genomes would deteriorate much faster into mutational meltdown, and thus error catastrophe (i.e., extinction) would occur much sooner!

This chapter has shown so far that the evolution problems created by (*i*) accumulation of nearly-neutral mutations, (*ii*) clustered mutations, (*iii*) natural selection limitations, (*iv*) biological noise, (*v*) selection thresholds for minor mutations, and (*vi*) selection interference provide insurmountable

evidence that the Primary Axiom is false, and should be rejected.

Creation of a New Gene

Approaching the crux of Primary Axiom truthfulness another way—by assuming it is valid and then analyzing what would be required to create a single new gene by natural selection—is highly enlightening. This analysis is provided here.

To begin, deleterious mutations will temporarily be ignored to determine if natural selection for beneficial mutations alone can do the job. Although clearly unrealistic, this scenario gives the Primary Axiom its best opportunity to succeed.

Step 1 — define a desirable mutation.

There are four types of nucleotide molecules in DNA, which are identified symbolically by the letters A (adenine), T (thymine), C (cytosine), and G (guanine). Each nucleotide will bond with only one other nucleotide type, as dictated by hydrogen bonding patterns, and become a *base pair*. The possible base pairs are G-C and A-T. By itself, no one nucleotide has more value than the others. Selection for a single nucleotide can only occur in the context of its surrounding nucleotides, so a nucleotide mutation can only be evaluated in the context of surrounding molecules. In other words, a mutation cannot be defined as good or bad, except in relation to its functionality that is shared with neighboring nucleotides.

The relationships between nucleotides and their surrounding neighbors illustrate the principle of *irreducible complexity* at its most fundamental biological level. This principle conveys the truth that a biological system composed of multiple interacting parts which contribute to its basic function will stop functioning if any one of the parts is removed from the system. The human eye is a prime example of an irreducibly complex system—each part of the eye, along with visual interpretation from the brain, is required to deliver the miracle of sight. Loss of any part, through injury for example, will result in blindness.

Because of irreducible complexity, natural selection must find itself in a paradox when attempting to create a new beneficial gene. To *create* a new function (i.e., create new biological information), natural selection will need to select for a first beneficial mutation. However, the value of that new nucleotide

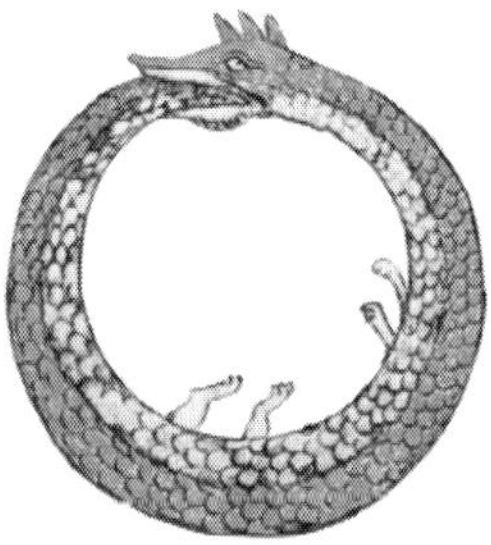

can only be defined in relation to its neighbors—and yet most of those neighbors will also have to change in order to create the new information. Thus, natural selection is actually in a circular paradox—it must destroy the very context upon which it is trying to build!

This real problem for natural selection—*epistasis,* or the fundamental interrelationship of nucleotides—is actually infinitely complex and virtually impossible to analyze. Therefore, it has been conveniently ignored by geneticists.

The human genome is literally a book, written in a language where short nucleotide sequences are actual sentences. Therefore, the possibility of random letters (nucleotides) accidentally falling into place and making a single meaningful sentence is numerically not feasible. The same is true for functional strings of nucleotides. Thus an unsolvable problem occurs in this very first step—trying to define what the first beneficial mutation should be!

Step 2 — wait for the first mutation.

Ignoring the unsolvable problem outlined above (as do other geneticists), the next step to create a new gene is to wait until the first mutation has occurred.

Human evolution is generally assumed to have occurred in a small population of about 10,000 individuals. Assuming 100 mutations per individual per generation (the same assumption used previously in this chapter), the mutation rate per individual per generation for any given nucleotide is very small—about one chance in 30 million (100 mutations divided by 3 billion nucleotides). Thus one would have to wait 3,000 generations (at least 60,000 years) for a specific nucleotide to mutate within a population of 10,000. Two out of three times, it will mutate into the wrong nucleotide (since the only base pair possibilities are G-C and A-T), so it will actually take twice as long (at least 120,000 years) to get the desired mutation in the correct site.

Once the mutation has occurred, it must become fixed (made permanent) in the population. This happens when all individuals within the population have two copies of the mutation.

Because new mutations are so rare, they have a very great probability of becoming lost from the population due to random genetic drift. Only if the mutation is dominant (does not require a copy from both parents to express itself)—and also has a very distinct benefit—does natural selection have a reasonable chance to rescue it from random elimination due to genetic drift. But almost all beneficial mutations are recessive (i.e., they require the same mutant gene to come from both parents in order to be expressed) and nearly-neutral (i.e., they are not selectable by natural selection). If a mutation increases fitness by 0.5%, it only has a 1% probability of becoming fixed within the population. Thus 99 times out of 100, the mutation will be lost through random genetic drift. A typical, mildly-beneficial mutation must therefore occur about 100 times before it is likely to become fixed within the population (even though it is beneficial!).

On average, it will take 12 million years (100 x 120,000 years) to fix the first typical beneficial nucleotide mutation in the population so it can become the basis for a new gene.

Step 3 — wait for other mutations.

The process explained above must now be repeated for all other nucleotides that will encode the new, hoped-for gene. Ignoring regulatory elements and introns (which are beyond the scope of this book), the smallest gene is minimally 1000 nucleotides long (50,000 nucleotides is a more realistic length). Assuming a linear process for the other 999 nucleotide mutations that are needed to create the new gene, it will take 12 billion years (12 million years x 1000) to create the smallest possible gene. This is approximately the amount of time since the big bang—which mainstream contemporary science claims was the beginning of the entire universe!

Step 4 — wait for recombination.

Since sexual species (including mankind) can shuffle mutations, one might believe that all the needed mutations for the new, tiny gene can occur simultaneously within different individuals within the population, and that they can then be spliced together via *recombination*. (Note: recombination is the process by which new combinations of genes are created within an individual to pass on to progeny; in other words, recombination is the process that ensures offspring are genetically different from each other.) Such splicing might eliminate the need for mutations to occur sequentially, thereby shortening the time required to create the new gene.

However, the probability of getting the required 1000 mutations to shuffle together in the hoped-for sequence is so exceedingly minute it would take **even more billions of years than the sequential approach!**

More importantly, geneticists who study the human genome have consistently seen that the genome is built from large blocks of nucleotides (20,000 to 40,000 nucleotides per block), and that essentially **no recombination within these blocks has occurred**

since the origin of man.[37] Thus, virtually no meaningful shuffling occurs at the level of individual nucleotides. Only large, gene-sized blocks of DNA are shuffled, which means that this step is not even a possibility.

Step 5 — wait on "Haldane's dilemma."

As discussed previously, the famous geneticist Haldane calculated that it would take six million years for 1000 mutations to become fixed within a population. Geneticists Crow, Kimura, and ReMine (as recently as 2005) have independently confirmed the accuracy of Haldane's calculations. To put Haldane's timeframe into perspective, six million years is the time it supposedly took for mankind to evolve from chimpanzees.

Haldane's timeframe to fix 1000 mutations (six million years) is 2000 times shorter than the timeframes developed in Steps 2 and 3 above. Nevertheless, Haldane's timeframe will be used in the analyses below in order to, once again, give the Primary Axiom its best opportunity to succeed.

One might jump to the conclusion from Haldane's calculations that one new, small gene of 1000 nucleotide mutations could actually be created in six million years. However, this is not a valid conclusion. Haldane's calculations only apply to *independent, unlinked* mutations. Natural selection for 1000 *specific* and *adjacent* mutations in one linkage block that would be needed to form such a gene could not happen in six million years—not even in six *billion* years! This is because the genome of a chimpanzee has billions of nucleotides, so the probability that 1000 random mutations would all occur in the same linkage block and become fixed in a population of 10,000 is infinitesimal. Remember that, according to Haldane, only 1000 mutations *total* could become fixed in a six million year period. To expect all 1000 to occur in a single linkage block is absurd.

The bottom line effect of Haldane's dilemma is that natural selection to fix new beneficial mutations occurs at glacial speeds—and its progress slows as the number of nucleotides under selection increases. Furthermore, since only 1000 *unlinked* beneficial mutations could have been fixed over the last six million years, the amount of new genetic information created would be miniscule—equivalent to merely 333 randomly chosen alphabetic characters. It would be impossible for such a small amount of new genetic information to transform a chimpanzee into a human being.

Step 6 — survive consecutive fitness valleys.

Evolutionists generally agree that creating a new gene results in a *fitness valley*. A partially-completed gene is neither beneficial nor neutral—it is always deleterious. Therefore, fixing a partially-completed gene in a species makes that species *less* fit. The period of time during which the species remains less fit is called a fitness valley.

One could assume that the species might survive fitness valleys if they were rare and relatively brief. However, lengthy fitness valleys are more likely to lead to extinction than forward evolution.

Because of Haldane's dilemma, the slowness of fixing beneficial mutations should extend the duration of fitness valleys indefinitely. Furthermore, if forward evolution were truly continuous, as widely claimed by mainstream science, all species would be continuously fighting for survival from consecutive fitness valleys. Life would always be under construction, so total fitness would always be declining rather than improving. As a result, extinction of species would be much more frequent than actually occurs in nature.

In the case of Step 5 above involving only 1000 mutations, consecutive fitness valleys would last a total of six million years!

Clearly, the concept of fitness valleys argues strongly against forward evolution.

Step 7 — overcome poly-constrained DNA.

Genetic science has discovered that most DNA sequences are *poly-functional,* which means they must also be *poly-constrained.* When a DNA sequence has meaning on several different levels (i.e., it is poly-functional), each level of meaning limits potential future change (i.e., the DNA sequence becomes poly-constrained).

To illustrate the meanings of *poly-functional* and *poly-constrained,* consider the palindrome:

A MAN, A PLAN, A CANAL—PANAMA.

Removing the spaces and punctuation from this sentence gives:

AMANAPLANACANALPANAMA

Reading this string of letters in reverse order gives exactly the same result:

AMANAPLANACANALPANAMA

This is a simple example of a poly-functional sentence. Although it only has two directions that can be read, it is poly-constrained because changing any letter will destroy the full meaning of *both* sentences, as illustrated here with three examples:

AMANA**X**LANACANALPANAMA

AMANAPLANA**Y**ANALPANAMA

AMANAPLANACANALP**Z**NAMA

Now consider this four level Latin palindrome, called the Sator Square:

S A T O R

A R E P O

T E N E T

O P E R A

R O T A S

This word puzzle dates back to at least AD 79, when it was covered by volcanic pyroclastic flows that smothered and destroyed the Roman city Herculaneum (located in the shadows of Mount Vesuvius). The puzzle consists of a sentence written in Latin—SATOR AREPO TENET OPERA ROTAS that can be read four different ways. The first letters of each word form the first word, the second letters form the second word, etc. Therefore, it can be arranged in a word square which reads four different ways—horizontally or vertically from either top left to bottom right or bottom right to top left. In English, the sentence essentially means, "The sower Arepo holds the work of the wheel."

As with the case of the two level palindrome A MAN, A PLAN, A CANAL—PANAMA, changing one letter in the Sator Square destroys the full meaning of the sentence at all four levels, as shown on the next page:

S A T O R

A R E P O

T T N E T

O P E R A

R O T A S

An example that more closely maps to a poly-functional DNA sequence is the poly-functional word LIVE. Read from the right, it spells EVIL. Thus two different words are communicated using the same letters. Changing any letter will destroy the intended meaning of both words, as shown in these examples—

HIVE, L**M**VE, LI**Q**E, LIV**G**.

All the examples shown above are poly-constrained, meaning that any changed letter will result in degeneration of meaning at all levels. Extending this concept to poly-functional DNA sequences, it should be obvious that **most nucleotide mutations can only degenerate or diminish genetic information—never improve or increase it!**

An article[38] written by Edward Trifonov was published in the *Bulletin of Mathematical Biology* in 1989 that was titled "The Multiple Codes of Nucleotide Sequencing." The abstract for this article contains the following statements—

"Nucleotide sequences carry genetic information of many different kinds, not just instructions for protein synthesis ... The codes are degenerate and corresponding messages are not only interspersed but actually overlap, so that some nucleotides belong to several messages simultaneously. Tandemly repeated [end-to-end repeated] *sequences frequently considered as functionless 'junk' are found to be grouped into certain classes of repeat unit lengths. This indicates some functional involvement of these sequences. A hypothesis is formulated according to which the tandem repeats are given the role of weak enhancer-silencers that modulate, in a copy number-dependent way, the expression of proximal genes. Fast amplification and elimination of the repeats provides an attractive mechanism of species adaptation to a rapidly changing environment."*

Edward Trifonov is a brilliant Russian-born Israeli molecular biophysicist who discovered the rules that determine the curvature of DNA molecules and their bending within nucleosomes. He also unveiled multiple codes in biological sequences as well as the modular structure of proteins. His article confirms the poly-functional nature of nucleotide sequences. Note that his last sentence above speaks to adaptive evolution, and that the third and fourth sentences negate the concept of "junk" DNA.

In 2003, an article[39] appeared in *Scientific American* titled "The Unseen Genome: Gems among the Junk." It was written by W. Wayt Gibbs and had the following first sentence—

"Just when scientists thought they had DNA almost figured out, they are discovering in chromosomes two vast, but largely hidden, layers of information that affect inheritance, development and disease."

Given the marvelous complexity of life, the amount of information packed into the human genome is absolutely incredible. Science does not yet know or understand all the levels of organization and encryption/data compression that are active in the genome. Nevertheless, the following poly-functional effects appear to be operative in addition to those cited above—

- Extensive sequence-dependent, 3D organization within chromosomes and the entire nucleus of cells.
- All DNA sequences encrypt up to *twelve* different codes.
- Every nucleotide interacts with many other nucleotides, and everything in the genome seems to be context-dependent.

As illustrated by the simple examples of poly-functional palindromes given above, every poly-functional DNA sequence is also poly-constrained. The existence of ubiquitous, genome-wide, poly-constrained DNA seems to be an overwhelming obstacle for forward evolution theory. Changing *anything* seems to potentially change *everything!*

Thus the poly-constrained nature of DNA serves as strong evidence that higher genomes cannot evolve via the Primary Axiom.

Step 8 — defy irreducible complexity.

In 1996, Michael J. Behe, an American biochemist, published a book[40] titled *Darwin's Black Box: The Biochemical Challenge to Evolution*. This book introduced the phrase *irreducible complexity* and applied it to evolution. The basic concept is this—in a complex system made of many parts, each part has limited value by itself, but has synergistic value within the context of the entire functional system; therefore, irreducibly complex systems cannot arise piecemeal by chance, but must come together all at once through intelligent design.

A mousetrap provides a simple example to illustrate the concept of irreducible complexity. The trap consists of five interacting parts—a wooden base and a spring, a catch, a hammer, and a hold-down bar. None of these parts has much functional value by itself, but integrated together into a mousetrap, they all work well together to reliably catch mice. If any part is removed, the mousetrap will no longer function.

Behe pointed out that biological systems also require multiple parts working together in order to function properly, and therefore are prime examples of irreducible complexity. He also asserted that natural selection cannot create such systems because they are not selectable until all the parts are already assembled and working together—i.e., natural selection cannot select for a function that is not already in place, nor can it create the function one piece at a time.

Of course, the concept of irreducible complexity is rejected by the scientific community at large, and is considered pseudoscience because it raises the need for intelligent design. One biologist, Kenneth Miller, belittled the concept by insisting that a mousetrap minus its hold-down bar and catch can be used as a spitball launcher or a tie clip, and that other parts can be used to create a clipboard, a key chain, or a toothpick. Notice that none of Miller's scornful examples have anything to do with catching mice, so they are completely irrelevant. Such is the ridicule that one challenging the Primary Axiom from common sense is likely to face from mainstream contemporary science.

Behe explained very clearly why irreducible complexity precludes forward evolution—

"An irreducibly complex system cannot be produced directly (that is, by continuously improving the initial function, which continues to work by the same mechanism) by slight, successive modifications of a precursor system, because any precursor to an irreducibly complex system that is missing a part is by definition nonfunctional."

Dr. Sanford is a strong supporter of irreducible complexity, as is the author of this book. Dr. Sanford brilliantly explained irreducible complexity this way—

*"It is in the realm of **mind** that deep complexity comes together and becomes integrated."*

Creating a functional gene from 1000 nucleotide mutations is much more complex than one might think before reading this from Dr. Sanford—

*"We do not yet really understand how any single gene from a higher life form really works—not in its entirety. Not in the context of everything else that is happening in the cell. A single gene with all its interactions is still way too complex for us. When we consider the full complexity of a gene, including its regulatory and architectural elements, a single gene has about 50,000 component parts. I presume that this is more component parts than are found in a modern automobile. There is no simple linear path that leads car components to spontaneously become a functional car—**mind** is obviously required*

(actually, ***many*** *brilliant minds). In the same way, there is no linear path of selection that can build a single gene from its individual nucleotides — a* ***mind*** *is likewise required.* ***Yet a single gene is just a microscopic speck of irreducible complexity, within the universe of irreducible complexity that comprises a single cell. Life is itself the very essence of irreducible complexity—which is why we cannot even begin to think of creating life ourselves. Life is layer upon layer upon layer of irreducible complexity.*** *... If we cannot do this—why would we think that random mutations, combined with a very limited amount of reproductive sieving, could accomplish this?"*

Step 9 — overcome being nearly-neutral.

Most geneticists have begrudgingly acknowledged the difficulty that natural selection has selecting against nearly-neutral deleterious mutations. However, Dr. Sanford has never heard the flip side of this evolution problem acknowledged—the fact that the obstacle of near-neutrality is much more severe for beneficial mutations than for deleterious mutations. Virtually *every* beneficial mutation must fall within Kimura's "no selection zone." This means that natural selection cannot select for *any* beneficial mutations. Thus a small gene consisting of 1000 beneficial mutations will never be selected and become fixed within a human population. Furthermore, there is no conceivable way to build a gene one nucleotide at a time using natural selection.

Step 10 — rise above millions of deleterious mutations.

As stated at the beginning of this exercise, deleterious mutations have been temporarily ignored to determine if natural selection for beneficial mutations alone can do the job. Since ignoring deleterious mutations is clearly unrealistic, they will now be included as the final step of this section because natural selection must operate in the real world—where deleterious nucleotide mutations outnumber beneficial mutations by *a million to one*.

As covered previously in this section, large blocks of DNA have no history of recombination, thus recombination occurs between genes rather than between nucleotides. Without recombination, such DNA blocks are subject to "Muller's ratchet." This means that, since beneficial and deleterious mutations within a block cannot be separated, blocks containing beneficial mutations will degenerate as multitudes of deleterious mutations drag the scarce, captive beneficial mutations down with them. Also, while waiting for the next rare beneficial mutation to occur, deleterious mutations will continue to deteriorate the DNA block. Furthermore, time (the enemy) causes some of the beneficial mutations to back-mutate into deleterious mutations.

Bottom line—**Muller's ratchet will destroy a gene long before it can form by natural selection.**

Biological Information—New Perspectives

In the spring of 2011, a landmark symposium was held at Cornell University in New York to review the latest scientific research on biological information. Twenty-four papers, written by twenty-nine different scientists, were presented. These authors represented the scientific disciplines of evolutionary theory, molecular biology, genetics, whole organism biology, developmental biology, physics, biophysics, mathematics, linguistics, information theory, computer science, numerical simulation, and thermodynamics. The proceedings of this symposium were published by World Scientific in 2013 under the title *Biological Information—New Perspectives*. A condensed, less technical synopsis of the papers (under the same title) is available free of charge at BINP.org.

The scientists who authored the papers generally agreed on the following three significant[47] points—

1. Information is the key to understanding life. Within the simplest cell there exists an immense flow of information through a mind-boggling system of information networks. There is constant and multidirectional communication between proteins, RNAs, and DNAs, and these biological information networks are in many ways similar to the Internet.

2. These biological information systems appear to greatly surpass human information technologies. Such information systems cannot possibly operate until all the countless components of the system are in place—including hardware, software, multiple languages, storage/transmission of communicable prescriptive information units, error testing/correction systems, designated senders/receivers, etc. Such systems must be comprehensive and coherently integrated before they can effectively operate.

3. The enormous amounts of information found within any cell, and the irreducibly complex nature of information systems, can no longer rationally be attributed to just the mutation/selection process. New perspectives are needed that might help us better understand the nature, origin, and maintenance of biological information.

It is clear from these three points of agreement that the symposium validated Dr. John Christopher Sanford's book *Genetic Entropy & The Mystery of the Genome*. Some examples of the research findings that go beyond what the author has already covered in this chapter include the following—

The Membrane Code: *cell membranes carry biological information that is necessary for embryo development but is not specified by DNA sequences. Membrane information is embodied in patterns that provide targets for the localization of intracellular molecules, global spatial coordinates in the form of electric fields, and an extracellular "sugar* *code" that mediates interactions with other cells.* ***Membrane patterns constitute a whole new layer of biological information, which cannot be reduced to DNA sequences or explained by DNA mutations, and thus cannot be understood in neo-Darwinian terms.***[48]

<u>Biocybernetics and Biosemiosis</u>: *the information networks found within living cells are remarkably similar to computer networks. Along with DNA, every protein and RNA simultaneously constitutes both hardware and software. As living algorithms, such information molecules simultaneously encode their own basic sequence, folding, transport, and biological function.* ***Nobody thinks that computer networks (including the associated hardware, software, language, and specified meaning), could ever arise spontaneously. So is it reasonable to think that vastly superior biological information systems, occurring just above the atomic level, could arise by any type of Darwinian trial/error process?*** [49]

<u>Limits of Chaos and Progress in Evolutionary Dynamics</u>: *selection is effective in optimizing a biological information system that is already in place, but it cannot lead to a continuous increase of new functions or new information. Instead, what was being selected for quickly reached a natural limit, at which point all advance stopped.* ***The evolutionary process driven by mutation-selection, in both mathematical models and directly observed behavior, is that of a system going to equilibrium and staying there.***[50]

<u>Multiple Overlapping Genetic Codes Profoundly Reduce the Probability of Beneficial Mutation</u>: *within biological information systems there are many overlapping messages and overlapping codes. This represents a very sophisticated form of data compression. Overlapping codes by themselves create major explanatory difficulties: How could they possibly arise by natural process alone? Moreover, because overlapping codes are poly-functional (one DNA sequence can code for two or more messages), they are also poly-constrained (each nucleotide contributes to more than one message—which greatly reduces* *the chance that any random change could be non-deleterious). The reality of overlapping codes drastically reduces the probability of unambiguously beneficial mutations.* ***This demands that scientists dramatically adjust downward their estimates of the actual rate of beneficial mutation. The actual rate could easily be 1000-fold less than has previously been thought.*** *Extremely low rates of beneficial mutation create very serious explanatory difficulties. If unambiguously beneficial mutations are extremely rare, then there are simply too few beneficial mutations to allow for genome-building. For example, the ape-to-man scenario requires that, minimally, tens of millions of beneficial mutations must arise and advance all the way to fixation in just 300,000 generations. This means that hundreds of beneficial mutations must be fixed every single generation.* ***The beneficial fixation rate would be many orders of magnitude too small to accomplish this.***[51]

<u>Getting There First: An Evolutionary Rate Advantage for Adaptive Loss-of-Function Mutations</u>: ***all "beneficial" bacterial mutations analyzed consistently turn out to entail a loss-of-function rather than creation of a new biological function.*** *Loss of function mutations involve such things as deleted genes, inactivated genes, or otherwise disrupted genes. This means that even useful adaptive mutations usually still involve a net loss of information. When a bacterial population encounters a new unfavorable environment, there is urgency in dealing with the new challenge (or else the population will go extinct).* ***This means that selection does not have time to find the best mutation for solving the problem, but will consistently amplify the first mutation that resolves the immediate need (even when the change is deleterious in the big picture).*** *Most adaptive selection events will amplify loss-of-function-mutations (they almost always arrive on the scene first).* ***Because loss-of-function mutations inherently involve loss of information, there tends to be a net loss of information even while meaningful adaptation is happening.***[52]

<u>Information Loss: Potential for Accelerating Natural Genetic Attenuation of RNA Viruses</u>: *loss of information is not always a bad thing. RNA viruses have a high rate of mutation, and are known to be prone to mutational degeneration. Erosion of information due to mutation accumulation probably plays a significant role in extinction of viral strains and the cessation of pandemics. Pharmaceuticals that increase RNA mutation rates should be highly effective in reducing severity and duration of pandemics. These theoretical studies have been subsequently validated by an historical analysis of the human H1N1 strain of influenza.* ***Since the pandemic of 1918, that influenza strain underwent continuous accumulation of mutations at a very constant rate, such that over 10% of its genome became mutated. During the same time, the strain underwent continuous and dramatic attenuation (as evidenced by reduced pathogenicity), and in 2009 the human H1N1 strain apparently went extinct.***[53]

<u>Information and Thermodynamics in Living Systems</u>: *what allows life to be alive is not biochemistry plus an external energy source. Rather, life is possible because of* **biological information***—which allows capture of external energy, directed energy processing and enhancement, and the use of such energy to build and run molecular machinery along with a multitude of metabolic networks. Such energy-rich molecular machines and networks are themselves specified and controlled by biological information.* ***The molecules, machines, and networks do not make the information. Rather, the information makes the molecules, machines, and networks. The neo-Darwinian view has always been just the opposite—organic molecules plus energy give rise to information. Information is primary, while molecules are secondary.***[54]

<u>Chapter Summary</u>

This chapter has covered the following critical points about the human genome—

1. There tends to be a net loss of genetic information even while adaptive evolution is occurring.
2. Most deleterious nucleotide mutations and virtually all beneficial mutations fall within Kimura's "no selection zone" and are therefore not subject to natural selection. Deleterious mutations outnumber beneficial mutations by a ratio of at least a million to one.
3. Nearly-neutral deleterious mutations accumulate, causing degeneration of the human genome.
4. Clusters of nucleotides form linkage blocks that mask beneficial mutations and never break apart. These linkage blocks are inherited as if a single trait, and have the net deleterious effect of the component mutations they encompass.
5. The average magnitude of harmful effect is typically greater for clustered mutations than for individual nucleotide mutations. Also, the frequency of clustered mutations is much higher than that of single mutations where the deleterious effects are most damaging.
6. Natural selection can select for or against minor mutations, which are more damaging than nearly-neutral mutations, but selection thresholds limit natural selection's effectiveness.
7. Biological noise severely limits the effectiveness of natural selection.
8. Selection interference severely limits the effectiveness of natural selection.
9. Natural selection does not have time to find the best mutation for solving a problem. Rather, it will amplify the first mutation that resolves the immediate need—even when the change is deleterious overall.
10. The evolutionary process driven by mutation and natural selection goes to equilibrium and stays there; it does not create new genetic information.

11. Cell membranes carry biological information that is necessary for embryo development but is not specified by DNA sequences. Membrane patterns constitute a whole new layer of biological information which cannot be reduced to DNA sequences or explained by DNA mutations.
12. The genetic molecules, machines, and networks that enable life do not make genetic information. Rather, genetic information makes the molecules, machines, and networks.

Taken together, these points provide comprehensive scientific proof that the Primary Axiom is false. Hence the theory of forward evolution is wrong.

Dr. Sanford wrote the following comments regarding his critical analysis of the Primary Axiom—

"We have reviewed compelling evidence that even when ignoring deleterious mutations, mutation/selection cannot create a single gene—not within the human evolutionary timescale. When deleterious mutations are factored back in, we see that mutation/selection cannot create a single gene—ever. This is overwhelming evidence against the Primary Axiom. ***In my opinion, this constitutes what is essentially a formal proof that the Primary Axiom is false.***

"In conclusion, the genome must have been designed, and could not have evolved. Yet we know that 'micro-evolution' (adaptive selection) does happen, correct? How can this be? … I believe most useful variation must be ***designed****. When we see adaptive selection occurring, we are usually witnessing segregation and recombination of useful variants of genes and gene components—****which were designed to segregate and recombine in the first place****. We are not usually seeing the result of random mutations—which are consistently deleterious. Selection operates to eliminate the worst of mutations, while favoring the most desirable recombinants and segregants of designed variation. … It is this type of designed*

diversity that natural selection can act upon most effectively. All such designed variants would be expected to be created within useful linkage groups, and would have originated at high allelic frequencies. ... Functional linkage groups and high allele frequency allow for very rapid responsiveness to selection, and thus rapid local adaptation."

The reader may be interested to know that Dr. Sanford is not the only scientist who believes that Darwinian evolution is a false theory. On February 20, 2006, the Discovery Institute[55] published the following article, written by Robert Crowther—

Over 500 Scientists Proclaim Their Doubts About Darwin's Theory of Evolution

Over 500 doctoral scientists have now signed a statement publicly expressing their skepticism about the contemporary theory of Darwinian evolution.

The *Scientific Dissent From Darwinism* statement reads: "We are skeptical of claims for the ability of random mutation and natural selection to account for the complexity of life. Careful examination of the evidence for Darwinian theory should be encouraged."

The list of 514 signatories includes member scientists from the prestigious US and Russian National Academy of Sciences. Signers include 154 biologists, the largest single scientific discipline represented on the list, as well as 76 chemists and 63 physicists. Signers hold doctorates in biological sciences, physics, chemistry, mathematics, medicine, computer science, and related disciplines. Many are professors or researchers at major universities and research institutions such as MIT, The Smithsonian, Cambridge University, UCLA, UC Berkeley, Princeton, the University of Pennsylvania, the Ohio State University, the University of Georgia, and the University of Washington.

Discovery Institute first published its *Scientific Dissent From Darwinism* list in 2001 to challenge false statements about Darwinian evolution made in promoting PBS's "Evolution" series. At the time it was claimed that "virtually every scientist in the world believes the theory to be true."

"Darwinists continue to claim that no serious scientists doubt the theory and yet here are 500 scientists who are willing to make public their skepticism about the theory," said Dr. John G. West, associate director of Discovery Institute's Center for Science & Culture. "Darwinist efforts to use the courts, the media and academic tenure committees to suppress dissent and stifle discussion are in fact fueling even more dissent and inspiring more scientists to ask to be added to the list."

According to West, it was the fast growing number of scientific dissenters which encouraged the Institute to launch a website—www.dissentfromdarwin.org—to give the list a permanent home. The website is the Institute's response to the demand for information and access to the list both by the public, and by scientists who want to add their name to the list.

"Darwin's theory of evolution is the great white elephant of contemporary thought," said Dr. David Berlinski, one of the original signers, a mathematician and philosopher of science with Discovery Institute's Center for Science and Culture (CSC). "It is large, almost completely useless, and the object of superstitious awe."

Other prominent signatories include U.S. National Academy of Sciences member Philip Skell; American Association for the Advancement of Science Fellow Lyle Jensen; evolutionary biologist and textbook author Stanley Salthe; Smithsonian Institution evolutionary biologist and a researcher at the National Institutes of Health's National Center for Biotechnology Information Richard von Sternberg; Editor of Rivista di Biologia / Biology Forum—the oldest still published biology journal in the world—Giuseppe Sermonti; and Russian Academy of Natural Sciences embryologist Lev Beloussov.

If you have a Ph.D. in engineering, mathematics, computer science, biology, chemistry, or one of the other natural sciences, and you agree with the following statement, "We are skeptical of claims for the ability of random mutation and natural selection to account for the complexity of life. Careful examination of the evidence for Darwinian theory should be encouraged," then please contact us at cscinfo@discovery.org.

The list of scientists holding doctorate degrees (either Ph.D. or M.D.) who have signed *A Scientific Dissent From Darwinism* was last updated in June 2015. At that time, 946 such scientists had signed the list.

Genetic Entropy

Dr. Sanford's honest and comprehensive scientific analyses of the Primary Axiom resulted in the *truth* that the human genome is deteriorating. In fact, the genomes of all living things must be deteriorating. While these truths may be startling to some readers, they are consistent with the laws and conditions of mortality that are covered in Chapter 1; specifically—

- All living things will eventually die.
- Natural processes always proceed in the direction of less order (or more randomness and entropy). A resulting consequence of this condition is that *everything* in mortality naturally degenerates; hence the cause of death.

The genome is a living thing, therefore it must follow the laws of mortality. Dr. Sanford refers to the degeneration of the genome as *genetic entropy*. This is a fitting term as it ties the Second Law of Thermodynamics with the living genome and the laws and conditions of mortality that the Lord instituted after the Fall of Adam and Eve. The author discussed entropy in Chapter 1 for natural processes. Clearly, degeneration of the human genome is a natural process that obeys the laws of mortality.

Dr. Sanford has pointed out that mainstream genetic science acknowledges the fact that the human genome is deteriorating, but does not accept the logical conclusion that the Primary Axiom is false. In the book titled *Science and Earth History—The Evolution/Creation Controversy,* author Arthur Strahler discusses entropy as it pertains to evolution. First he acknowledges that creationists claim evolution runs counter to the Second Law of Thermodynamics and therefore living forms cannot evolve from simpler to more complex forms. Next he puts a spin on the Second Law of Thermodynamics by stating that the only truly closed system is the entire universe!

(Remember that the Second Law of Thermodynamics applies to closed systems that do not have external energy coming into them). Referring to the Chapter 1 example of ice melting according to the Second Law of Thermodynamics, one would conclude from Strahler's definition of a closed system that the ice should not melt because it is not in a closed system when placed on a table—but the ice *does* melt! Hence anyone with common sense can see that Strahler's attempt to distance the Second Law of Thermodynamics from evolution does not reflect reality.

Strahler and other mainstream scientists position the entire earth as an open system, with energy coming from the sun to fuel natural processes that, on a molecular level, can result in a decrease in entropy instead of the increase in entropy that the Second Law of Thermodynamics requires in closed systems. These scientists[41] emphasize—

"... the ability of living organisms as open systems to place energy temporarily in storage in complex molecules and to release energy as required by breakdown of those molecules. ... this localized decrease in entropy in no way violates the laws of thermodynamics, because the conversion and storage processes require the expenditure of even more energy as waste heat than the quantity of energy that is chemically stored."

Another example that is used to justify why the Second Law of Thermodynamics does not apply to biological systems on a molecular level is the hydraulic ram, or hydram. This physical device is powered by the kinetic energy of flowing water. Without requiring any other external energy, the hydram takes in water at one pressure and flow rate and converts it into a higher pressure with less flow. Although the hydraulic ram is not a biological system, it is touted as an example of how the Second Law of Thermodynamics can be waved aside in biological systems.

Thus these scientists have convinced themselves that biological processes are essentially not subject to the laws of mortality. However, their arguments are flawed, as Dr. John C. Sanford has revealed in his book on genetic entropy and Dr. Andy C. McIntosh has detailed in his paper[54] *Information and Thermodynamics in Living Systems*.

Incidentally, this same argument is used by these scientists to explain how life came about spontaneously from non-living, self-replicating molecules via *abiogenesis* (or *biopoiesis*). Surprisingly, however, author Arthur Strahler opens the possibility of "a Divine Foot in the door" for abiogenesis[41] in his book, but only to those who accept the Primary Axiom—

"But consider, then, that there is also another and later point of ontological convergence: the question of origin of living matter on earth. For both theists and deists who subscribe to the geological time scale in which the initial appearance of life goes back to something on the order of 3.5 billion years, or perhaps more, and who also consider organic evolution by natural selection to be a strong working hypothesis, it is permissible to hold the belief that God created the initial living substance. For the mechanistic materialists, it is permissible to postulate a naturalistic, spontaneous mode of biopoesis (origin of living matter) based on what is known today about chemical and physical processes and materials. In other words, this is one of the singular points in universal history when both dualistic and monistic ontologies come up with exactly the same result, which is the unique 'seed' or 'seedling' of the tree of life. Whichever origin is preferred, one being a religious belief and the other a naturalistic hypothesis of science, the question of continuous evolution of life thereafter can be examined in full independence of the question of its origin."

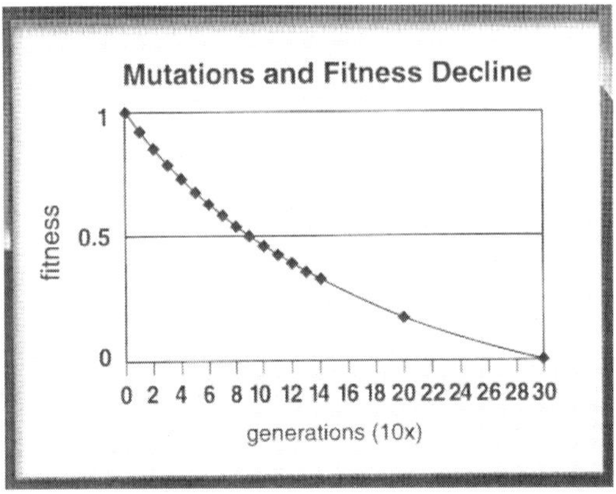

Figure 7

For the author of this book, the strongest evidence of the reality of genetic entropy is gleaned by plotting the lifetimes of early mankind, as recorded in the Bible, and comparing them to the simulation of Crow's model of fitness decline, as provided in the two graphs shown here and on the next page. Figure 7 is the simulation of Crow's model that was discussed in the section above titled Evaluating Crow's Solution. Remember that, even though Crow used both unrealistic and unreasonable assumptions in his model, the simulation still shows a decline in fitness over time that leads to eventual human extinction beginning in about 300 generations (6000 years).

Figure 8 (see the next page) was developed by Dr. Sanford. He plotted ages recorded in the Bible starting with Shem (one of Noah's sons). Notice the decline in fitness that Dr. Sanford's graph shows in comparison with the Crow model. Both illustrate the marked decline in fitness that mankind has experienced since the Flood. Both are also consistent with the concept of genome degeneration caused by mutation accumulation.

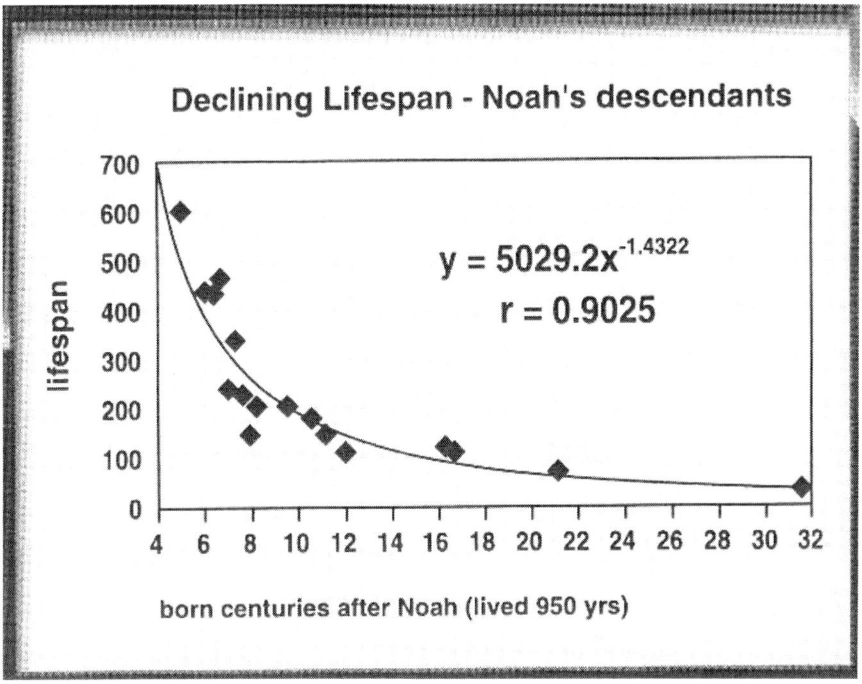

Figure 8

In Figure 8, the four points plotted between 9 and 12 centuries after Noah are those of Abraham, Isaac, Jacob, and Joseph. When Joseph introduced his father Jacob to the Pharaoh of Egypt during the seven-year famine, Pharaoh asked Jacob his age. Interestingly, Jacob's response validates what is shown in these two figures—

And Jacob said unto Pharaoh, The days of the years of my pilgrimage are an hundred and thirty years: few and evil have the days of the years of my life been, and have not attained unto the days of the years of the life of my fathers in the days of their pilgrimage. (Genesis 47:9)

The Bible is not the only record of mankind that tells of people living for hundreds of years. Similar claims of long life spans are found in the secular literature of several ancient cultures, including the Babylonians, Greeks, Romans, Indians, and Chinese.[43]

The Sumerian King List[44] is an example of such a record. The King List is an ancient manuscript originally recorded in the Sumerian language that lists the kings of Sumer (ancient southern Iraq), their length of reign, and their place of residence. This early manuscript alludes to a Flood with the statement, *"Then the flood swept over."* The King List records a marked decrease in life span ranging from the antediluvian kings to the postdiluvian kings. Although the life spans of the antediluvian kings are roughly sixty times longer than those of Biblical patriarchs, a paper[45] written by J. Walton in 1981 proposes that the difference can be explained by considering the numerical systems assumed (i.e., sexigesimal (base 60) for the King List and decimal (base 10) for the Bible).

Genetic Entropy Consequences

Dr. Sanford's analyses of the Primary Axiom have shown that nothing can stop degeneration of the human genome. What does this mean for humanity?

During the 1970s, geneticists believed that the forward evolution of mankind would be able to reverse the effects of genome deterioration. Some even had grandiose visions of what could be accomplished. Note the words of Kimura[46] in 1976 (highlights are the author's, not Kimura's)—

"Shall we be content to preserve ourselves as a superb example of living fossils on this tiny speck of the universe? Or, shall we try with all our might, to improve ourselves to become supermen, and to still higher forms, to expand into the wider part of the universe, and to show that ***life after all is not a meaningless episode?****"*

The author finds it disturbing that such a prominent geneticist considered his own life to be a *meaningless episode*—and yet, this is exactly what mainstream contemporary science teaches mankind today!

Earlier in his career, Dr. Sanford had similar thoughts, but he now understands they were based on a false hope. However, he does have hope for mankind that he shares in his book—

"Even as we cannot create life, we cannot defeat death. Yet I assert that there is One who ***did*** *create life, and who designed the genome. I do not know how He did it, but somehow He surely made the hardware, and surely must have written the original software. He is called the Author of Life (Acts 3:15—NIV). I believe the Author of Life has the power to defeat death and degeneration. I believe this is the* ***good news****.*

"It is my personal belief that Jesus is our hope. I believe that apart from Him there is no hope. He gave us life in the first place—so He can give us new life today. He made heaven and earth in the first place—so He can make a ***new*** *heaven and earth in the future. Because He rose from the dead—we can be raised from death, even that death which is already enveloping us. In these profound yet simple truths, I believe there is a true hope. I believe this is the hope which is unshakable, because I believe it is founded on the One who is eternal. It is a hope that has withstood the attacks of time, and the corruption of religion. It is a hope freely given, to anyone who would receive it today. I humbly put before you this alternative paradigm for your consideration—Jesus is our one true hope."*

The author commends Dr. Sanford for his faith in the Lord Jesus Christ and in the resurrection. He also congratulates Dr. Sanford for his willingness to point his readers to religious truths rather than scientific hypotheses in this very important matter concerning the future of mankind.

The author believes that God never intended for mankind to become extinct on earth, for His work and His glory are to bring to pass the immortality and eternal life of man. As the Apostle Peter wrote—

Nevertheless we, according to his promise, look for new heavens and a new earth, wherein dwelleth righteousness. (2 Peter 3:13)

The bottom line of genetic entropy is that mankind need not fear degeneration of the genome—God intended for it to degenerate, along with all else that is mortal, and He will ensure that all mankind will be resurrected with immortal bodies. In addition, He will create a new, immortal earth. Truly, this is wonderful news!

Chapter Conclusion

The author believes that Dr. John C. Sanford was raised up by God to provide scientific proof that man was not a development from lower orders of the animal kingdom. Since Dr. Sanford's work persuades mankind to believe in Christ, it is clearly of God, and should be welcomed with open arms by all Christians.

The author has read materials written by Christian scientists who seem puzzled why so many Christians have resisted the theories of modern Darwinian evolution. Evidently, these scientists have never experienced that uncomfortable dilemma which has plagued so many Christians as they study the tenets of contemporary science. One such writer positioned Charles Darwin as an instrument in God's hands for bringing forth truth.

The author of this book has a different view of Darwin, as does author Richard Webster, who argued that Darwin's *The Descent of Man* was influenced by racial prejudice, and that in it Darwin looked forward to the extermination of what he considered to be savage races.[30]

Darwin may also be considered the father of eugenics, since he repeatedly pointed to human efforts in animal and plant breeding as a model for man-directed selection to genetically improve the human race. Ever since Darwin, his followers have been eugenicists advocating the genetic improvement of the race. Before World War II, many nations, including the United States of America, had government-directed eugenics programs. These governmental programs included forced sterilization of the "unfit" and aggressive promotion of abortion and fertility control for the underclasses. After the horrors of the World War II holocaust, open discussions of eugenics were quietly put aside,[56] but only temporarily. Unfortunately, eugenics is now in favor again with elitists. The author believes eugenics is the primary motive behind forcing all Americans to pay for abortions and free contraceptives via the Affordable Care Act of 2010. Consequently, the author encourages the reader to decide for himself/herself if Darwin was an instrument in the hands of God.

Chapter Three

The Geologic Record

Contrary to popular belief, the geologic record does not support the theory of Darwinian evolution. In fact, breakthrough science is challenging old thinking about the geologic record!

The Geologic Record According to Mainstream Science

Prior to the 1770s, most scientists believed that the Great Flood of Noah's day was responsible for the geologic record, which consists of various layers of fossil-bearing sedimentary rock that extend over extremely wide areas—some nearly continental in extent. During that time, it was believed that all rocks had precipitated (i.e., been deposited) from the waters of a single, enormous Flood.[1] The fundamental assumption behind this early science was that the Flood was a global event which left its stratified signature on the earth's crust.

Sedimentary rock strata in northern Colorado

In 1785, James Hutton, a Scottish geologist and naturalist, read his entirely new theory about rock formation to members of the Royal Society of Edinburgh. He proposed that (*i*) the interior of the Earth is hot, and that its heat is the source of creating new rock; (*ii*) land is eroded by air and water and deposited as layers of sediment in the sea; and (*iii*) the earth's internal heat transforms the sediment into stone and then uplifts it to form new lands. Hutton also introduced the idea that the earth is very old, and that its history extends indefinitely into the past. His theory was based on the concept that *"from what has actually been, we have data for concluding with regard to that which is to happen thereafter."*

In 1788, Hutton's hypotheses were published as "Theory of the Earth; or an Investigation of the Laws Observable in the Composition, Dissolution, and Restoration of Land upon the Globe." The now celebrated conclusion of this paper was,

"The result, therefore, of our present enquiry is, that we find no vestige of a beginning—no prospect of an end."

Following criticism that his theory was atheistic and illogical, Hutton published an extended version of the theory as a two volume work in 1795. His work introduced the concept of *uniformitarianism*, which means that what science sees happening in nature today applies to the distant past as well as the future. Hutton also advocated uniformitarianism for living things, foreshadowing evolution, and even suggested that natural selection might be the mechanism making it happen—

"... if an organised body is not in the situation and circumstances best adapted to its sustenance and propagation, then, in conceiving an indefinite variety among the individuals of that species, we must be assured, that, on the one hand, those which depart most from the best adapted constitution, will be the most liable to perish, while, on the other hand, those organised bodies, which most approach to the best constitution for the present circumstances, will be best adapted to continue, in preserving themselves and multiplying the individuals of their race."

Today, James Hutton is considered the originator of uniformitarianism and the father of modern geology.[2]

In 1830, the foremost geologist of his time—Sir Charles Lyell—published the first volume of *Principles of Geology*, a book that endorsed Hutton's uniformitarianism and endeared it to modern science with the phrase—*"The present is the key to the past."* Lyell was the first to believe that the earth is more than 300 million years old, and was an influential friend of Charles Darwin. Darwin carried the first volume of *Principles of Geology* with him on the voyage of the *HMS Beagle* to Argentina and the Galapagos Islands that made him famous. During that voyage, Darwin considered himself a geologist and interpreted what he saw geologically "through Lyell's eyes." After the voyage, Lyell invited Darwin to dinner and the two became close friends thereafter. Ironically, Lyell, who was a devout Christian, had difficulty reconciling his personal beliefs with Darwin's *On the Origin of Species by Means of Natural Selection*.[3]

Today, modern science considers the concept of uniformitarianism to be a fundamental principle that supports evolution on earth and applies throughout the entire universe. As the geomorphology scientist, Dr. Arthur N. Strahler, wrote[4] (highlights are the author's, not Strahler's)—

"Uniformitarianism should be viewed as antithetic to any and all theories invoking supernatural forces for the origins of things geologic or biologic or of any other natural-science phenomena. Uniformitarianism includes the assumption that ***laws of physics and chemistry as presently understood have applied to all past time and will apply to all future time****; they are assumed valid unless compelling independent evidence should be found to suggest that they may have differed in quality or quantity at other points in time or space. In using that assumption,* ***all science operates today under the uniformitarian principle****. In cosmology, the study of the structure and evolution of the universe,* ***it is assumed that the laws of physics are similar throughout the entire universe****."*

A high level summary of the geologic time scale that is embraced by mainstream contemporary science—also by the United States Geological Survey and by the International Commission on Stratigraphy—is provided in Figure 9 below. Notice that the age of the earth, as shown in the clock just right of the 12 o'clock position, is 4.55 billion years (4550 million years). Also, life began about 4 billion years ago. The fossil record did not begin until near the end of the Hadean *eon,* so the other eons—Archean, Proterozoic, and Phanerozoic (which is not shown)—encompass the evolution of life. Together, the Paleozoic, Mesozoic, and Cenozoic *eras* form the Phanerozoic eon. A two million year Quaternary *period* (not shown), which is the time that recognizable humans have been on the earth, is too small to be visible at the scale of this clock.[5]

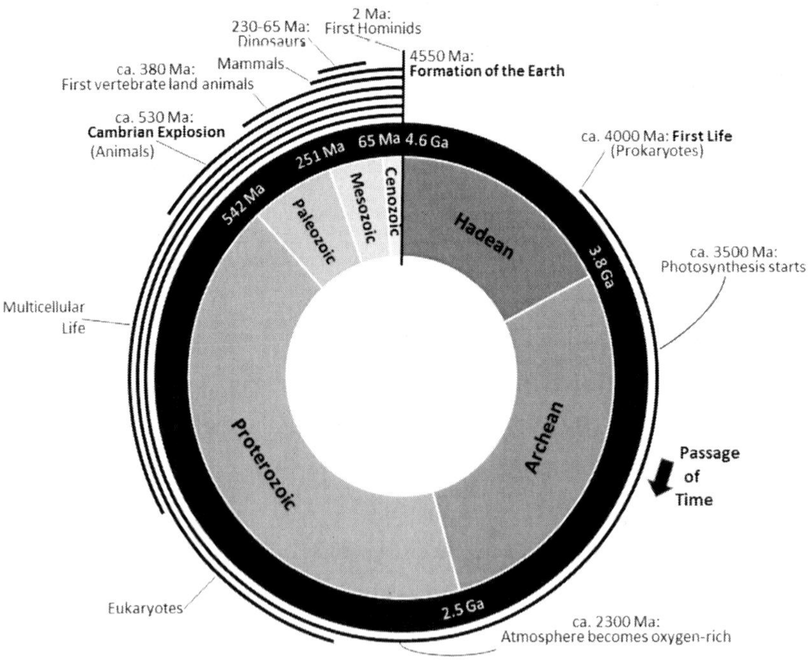

Figure 9—Geologic Time Scale Shown as a Clock
(Ma = 1 million years, Ga = 1 billion years)

The geologic time scale in Figure 9 is organized to correspond with changes in the composition of the earth's strata (i.e., layers of rock with internally unique but consistent characteristics). For example, rocks corresponding to the Hadean eon contain no fossils; however, fossils are found in the rock layers directly above the rocks in the Hadean layer. The circular lines that partially circumscribe the geologic clock in Figure 9 indicate how long different types of living things existed on the earth, as revealed in the fossil record. For example, dinosaur fossils are found in the strata corresponding to the Mesozoic era. Radioactive dating methods are used to determine the age of rocks, fossils, archeological artifacts, and the earth itself in order to create this geologic time scale.

The reader may find it interesting that the Apostle Peter prophesied about uniformitarianism in the last days (highlights are the author's)—

Knowing this first, that there shall come in the last days scoffers, walking after their own lusts, And saying, Where is the promise of his coming? for since the fathers fell asleep, ***all things continue as they were from the beginning of the creation.*** (2 Peter 3:3-4)

Accuracy of Geologic Dating Methods

As mentioned above, radioactive dating is used to determine the age of rocks, fossils, archeological artifacts, and the earth itself. Radioactive dating methods compare the amount of a naturally occurring radioactive material in a rock sample, or artifact, with the amount of its decay material that is also found in the sample. The comparison utilizes the measured decay rate of the radioactive material.

Radioactive decay occurs when the nucleus of an unstable atom emits particles. After the particles have been emitted, the atom becomes a different element because it has a different nucleus, and thus takes on the characteristics of the new element.

This is called *nuclear transmutation*. For example, a radioactive uranium atom emits protons and neutrons from its nucleus, the loss of which ultimately transmutes the atom into a lead atom, which is not radioactive. Uranium-lead dating is used for dating materials that are assumed to be extremely old (i.e., millions to billions of years). As another example, the radioactive version of a carbon atom (which has two more neutrons in its nucleus than a nonradioactive carbon atom) is transmuted into a normal carbon atom as it decays. Radioactive carbon dating is assumed to be accurate for materials that are less than 90,000 years old.

In order to apply radioactive dating methods, certain assumptions must be made. *The accuracy of these assumptions has a direct impact on the accuracy of the dating method.* Three assumptions that must be made are—

1. The decay rate of the radioactive material over time.
2. How much of the decay material existed in the sample to be dated when it was first created.
3. How much of the radioactive and decay materials have been added to, or taken away from, the sample over time.

Unfortunately, it is never possible to ensure accurate assumptions because of the simple fact that no scientist was available to observe what happened to the sample over time! Scientists can only observe what is happening now—certainly not 90,000 or 4.5 billion years ago.

This is where the principle of uniformitarianism comes into play for modern science—and it clearly influences the results of dating techniques. The assumptions that are used for radioactive dating by contemporary mainstream scientists, which are based exclusively on uniformitarianism, are as follows—

1. Today's measured decay rate of the radioactive material has not changed over time.

2. The amount of decay material that existed in the sample when it was first created is zero.

3. No radioactive or decay materials were added to, or taken away from, the sample over time.

New science has shown that these assumptions result in unreliable dating!

For example, a paper[10] published in 2004 about radioisotopes and the age of the earth discusses the findings of Dr. D. Russell Humphreys and three colleagues regarding a helium diffusion study of microscopic zircon crystals that were recovered from a borehole in Precambrian granitic rock in New Mexico. The Precambrian *super eon* covers the time span of the Hadean, Archean, and Proterozoic eons identified in Figure 9 (a span of about four billion years). Using uranium-lead dating techniques, samples of the granite rock were dated at 1.5 billion years, which puts them approximately in the middle of the Proterozoic eon.

The radioactive decay process that transmutes uranium into lead also produces atoms of helium, and the rate of diffusion (or leakage) of these helium atoms within the zircon crystals can be precisely measured. Some core samples of the Precambrian granite rock from New Mexico were given to the Oak Ridge National Laboratory (a high precision dating laboratory) for analysis. *The results showed that the measured rate of helium diffusion from the zircon crystals was compatible with the crystals being about 5,000 years old—not 1.5 billion years old!*

Why such an astonishingly large difference in dating results?

Another recent finding also throws doubt on uniformitarianism dating methods. On May 18, 1980, a volcano in the State of Washington named Mount St. Helens erupted. An extremely powerful explosion devastated its northern slope, and a new lava dome formed, shown here in Figure 10. This dome consisted of an igneous rock called dacite. Because the dome was formed by molten lava soon after the explosion occurred, radioactive dating should have resulted in a reading of "too young to measure." *However, when the potassium-argon (K-Ar) dating method was employed, it gave an unexpected reading of 2.4 million years!* [11]

Figure 10—Mount St. Helens Lava Dome
courtesy U.S. Geological Survey, Wikimedia Commons (in the public domain)

According[12] to Dr. Brent Dalrymple, a 2003 recipient of the United States National Medal of Science—

"The K-Ar method is the only decay scheme that can be used with little or no concern for the initial presence of the daughter isotope. This is because ^{40}Ar is an inert gas that does not combine chemically with any other element and so escapes easily from rocks when they are heated. Thus, while a rock is molten, the ^{40}Ar formed by the decay of ^{40}K escapes from the liquid."

However, Dr. Andrew Snelling, an Australian geologist, challenges what Dr. Dalrymple wrote above. Said Dr. Snelling in a recent[13] article—

"... this dogmatic statement is inconsistent with even Dalrymple's own work 25 years earlier on 26 historic, subaerial lava flows, 20% of

which he found had non-zero concentrations of ^{40}Ar (or excess argon) in violation of this key assumption of the K-Ar dating method."

Remember that a key, standard uniformitarianism assumption about radioactive dating is that the amount of decay material that existed in a sample when it was first created is zero. If argon gas (^{40}Ar) is present when rock is first formed from solidifying molten lava, the potassium-argon (K-Ar) dating method will consistently produce excessively old age-readings. While claiming such excessively old age-readings may be desirable for modern science in order to support Darwinian evolution, the claims appear to be pseudoscience!

Consider the following incomplete list of known locations where historical volcano lava flows were dated millions of years older than their actual ages—

Location / Year	**Rock Type**	**Actual Age in Years**	**Radioactive Age-Reading in Years**
Mt. Etna, Sicily / 1964	Basalt	50	.7 million
Akka Water Fall, Hawaii	Pleistocene	2100	32.3 million
Kilauea Iki, Hawaii / 1959	Basalt	55	8.5 million
Mt. Stromboli, Italy / 1963	Basaltic andesite	51	2.4 million
Medicine Lake Highlands, Glass Mountains, California	Obsidian	< 500	12.6 million
Rangitoto, Auckland, New Zealand	Basalt	< 800	.15 million

Location / Year	**Rock Type**	**Actual Age in Years**	**Radioactive Age-Reading in Years**
Mt. Erebus, Antarctica / 1984	Anorthoclase	30	.64 million
Kilauea, Hawaii	Basalt	< 200	21 million
Kilauea, Hawaii	Basalt	< 1000	42.9 million
Mt. Ngauruhoe, New Zealand / 1954	Andesite	60	3.5 million

Based on this information, Dr. Snelling wrote at the end of his article (highlights are the author's, not Dr. Snelling's)—

"... when samples of volcanic rocks are analyzed for K-Ar and Ar-Ar 'dating,' the investigators can never really be sure therefore that whatever ^{40}Ar is in the rocks is from in situ radioactive decay of ^{40}K since their formation, or whether some or all of it came from the mantle with the magmas. ... Furthermore, there would be no way of knowing, because the ^{40}Ar from radioactive decay of ^{40}K cannot be distinguished analytically from primordial ^{40}Ar not from radioactive decay ... ***Thus all K-Ar and Ar-Ar 'dates' of volcanic rocks are questionable, as well as fossil 'dates' calibrated by them."***

Another revealing example of new science that exposes the unreliability of radioactive dating involves the use of advanced radiocarbon dating.[14, 15] As explained above, the radioactive version of carbon atoms (^{14}C) has two more neutrons in its nucleus than the nonradioactive version (^{12}C or just C). Radiocarbon (^{14}C) is formed when cosmic rays interact with nitrogen in the atmosphere. The radiocarbon atoms combine with oxygen to form radioactive carbon dioxide, which is then assimilated by plants through photosynthesis. As animals eat the plants, they ingest the radiocarbon. Carnivorous animals consume radiocarbon by eating herbivorous animals. Assimilation of radiocarbon stops when the plants and animals die. Over time, the radiocarbon atoms inside the organisms are transmuted at a known rate as they decay into normal carbon atoms; hence, measuring the ratio of radioactive to nonradioactive carbon atoms ($^{14}C/C$) in a fossil sample provides a means for calculating the age of the fossil.

About twenty-seven years ago, a new technology was developed that more precisely measures this ratio. The technology utilizes an ion beam accelerator in conjunction with a mass spectrometer and is known as the AMS (accelerator mass spectrometer) dating method. Before the AMS method was available, the sensitivity of carbon dating was limited to a radioactive to nonradioactive ratio ($^{14}C/C$) no smaller than 1% of the modern atmospheric $^{14}C/C$ ratio value (this limit is notated as 1 *percent modern carbon*, or 1 *pmc*). The AMS method improved the sensitivity to 0.001% of the modern value (i.e., 0.001 pmc). This 1000-fold improvement extended the theoretical accuracy range for measuring a sample's age from about 40,000 years to 90,000 years.[16]

Scientists expected that the improved precision of the AMS method would result in new dating of dramatically older fossil samples. **However, no fossil samples have been found anywhere in the world that have a $^{14}C/C$ ratio as low as 0.001%, and therefore they cannot be as old as 90,000 years!**

Since most scientists involved with the AMS dating method believe the standard geologic time scale (Figure 9) is accurate, they naturally do not accept the results of AMS dating at face value. Instead, they insist that some type of contamination must have occurred in the samples that were tested. A 2001 paper[17] written by Paul Giem addresses this issue. Giem researched seventy AMS measurements of organic samples from the Phanerozoic eon (up to 542 million years old) that *should* have had zero radioactive carbon content (since all ^{14}C atoms should have decayed in the samples after 250,000 years because they have a half-life of only 5730 years—which means the number of ^{14}C atoms in a sample is reduced by half every 5730 years). Giem found that samples from every portion of the Phanerozoic eon showed detectable amounts of ^{14}C. For those AMS measurements considered the most reliable, $^{14}C/C$ ratios ranged from 0.1 to 0.5 pmc. After considering all possible sources of contamination, Giem concluded that the radioactive carbon detected in the samples likely came from the organisms from which the samples were derived, and therefore the AMS readings were accurate. Perhaps even more significant, since most fossil carbon seems to have approximately the same $^{14}C/C$ ratio, Giem concluded **it is plausible that all the fossilized organisms lived on the earth at the same time!**

Paul Giem is not the only scientist to report such results from AMS carbon dating. A paper[18] published in 2001 by six authors from the Leibnitz Laboratory at Christian-Albrechts University in Kiel, Germany reported detailed studies of mussel and snail shells obtained from a site in northwestern Germany. These shells had been dated using the Uranium/Thorium method as being 120,000 years old. AMS carbon dating results ranged from 0.1 to 0.5 pmc (i.e., 45,000 to 58,000 years old). The team also dated shells of other types of aquatic organisms which came from the northwest coast of Africa and were previously dated at 455,000 years. **AMS readings from 115 separate measurements averaged a remarkably high reading of 0.23 pmc, which corresponds to 50,000 years—not 455,000 years!**

Other papers report similar results, but none offer an explanation as to why carbon dating does not agree with other forms of radioactive dating methods. This is a huge, unanswered stumbling block for mainstream contemporary science!

There are two other aspects of carbon dating that must also be considered. As mentioned above, the carbon dating method compares a fossil sample's measured radioactive to nonradioactive ratio ($^{14}C/C$) to the *modern atmospheric* $^{14}C/C$ *ratio value*. Mainstream science, clinging to uniformitarianism, *assumes* that the modern atmospheric $^{14}C/C$ ratio value has been essentially constant since the Phanerozoic eon began 542 million years ago, which implies that the rate of ^{14}C production in the atmosphere equaled the rate of ^{14}C decay during this time. In other words, ^{14}C production and decay are in equilibrium today and have been for hundreds of millions of years. However, recent meticulous measurements[70] have revealed that the rates of ^{14}C production and decay are *not* in equilibrium! The rate of ^{14}C production actually exceeds the rate of ^{14}C decay by 30%—which means there was less ^{14}C in the atmosphere when an ancient animal died than there is in the atmosphere today. Since ^{14}C production surpasses ^{14}C decay by a whopping 30%, the amount of ^{14}C in the atmosphere has been increasing over time. The bottom line conclusion of this breakthrough in carbon dating science is that any calculated age based on the uniformitarianism assumption is too old!

Furthermore, several papers[17, 78, 79] have compared the amount of carbon in earth's biosphere before the Flood with the amount of carbon that exists in today's biosphere. As will be explained later, the earth's pre-Flood climate was subtropical over all its land areas, resulting in the natural growth of significantly more vegetation than naturally thrives today. During the Flood, this extensive vegetation became the source of coal, oil, and oil shale that supply much of the world's energy needs today. The papers referenced above estimate that the amount of carbon residing in the abundant plant life before the Flood was at least 100 times greater than is found in today's vegetation population. However, the amount of ^{14}C radiocarbon in the atmosphere before the Flood was likely *even less* than is calculated using the 30% growth rate factor explained above. The earth's pre-Flood magnetic field was significantly stronger than the field is presently (this will be covered in more detail later). A more powerful magnetic field would reduce the density of cosmic rays that could penetrate the magnetic field to interact with atmospheric nitrogen and create ^{14}C. As a result, the $^{14}C/C$ ratio value before the Flood was on the order of 1 pmc (i.e., 1 percent modern carbon)—which is 100 times lower than today's modern carbon ratio!

Taking into account (*i*) an initial $^{14}C/C$ ratio value two orders of magnitude lower than today's modern carbon value and (*ii*) a $^{14}C/C$ ratio value that is growing by 30% every year, there is no way that carbon dating using uniformitarianism assumptions can result in accurate results! Using the breakthrough science described in the two paragraphs above, the authors of the paper[15] introducing this science concluded (highlights are the author's, not the writers of the paper)—

"When we combine all these factors, we conclude it is not at all surprising organic materials buried in the cataclysm [Flood] *should display the roughly 0.05—0.5 pmc we actually observe.* ***We note that when these considerations are included, especially the larger pre-cataclysm carbon inventory, a $^{14}C/C$ value of 0.24 pmc, for example, is consistent with an actual age of 5,000 years."***

Unreliable Carbon Dating Example

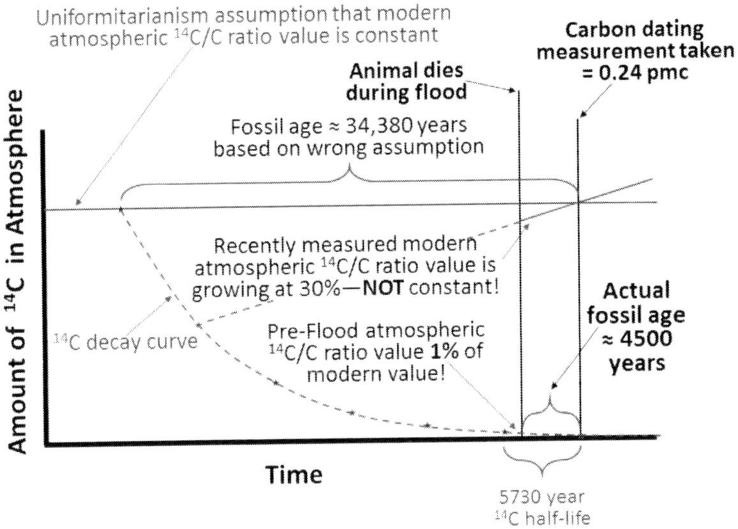

The graph above illustrates visually why a radiocarbon-dated fossil age using uniformitarianism assumptions is too old. The vertical scale shows the amount of ^{14}C in the atmosphere while the horizontal scale depicts the passage of time. When the animal died (during the Flood), there was substantially (at least 99%) less ^{14}C in the atmosphere than uniformitarianism assumes, and therefore the animal had considerably less ^{14}C in its body than assumed. The dotted line shows the ^{14}C decay curve, which, as explained above, has a half-life of 5730 years. Each star on the decay curve represents the passing of another 5730 years. The top horizontal line represents a constant $^{14}C/C$ ratio value which pinpoints when the animal would have died under the assumption that the modern atmospheric $^{14}C/C$ ratio value is a constant. But in reality, the animal died much more recently, having far less ^{14}C in its body at the time it died than assumed. This initial amount is lower than assumed because recent measurements show the $^{14}C/C$ ratio value is growing by 30% a year, and also the pre-Flood $^{14}C/C$ ratio value was no more than 1 pmc (i.e., 100 times lower than currently measured). Taking into account the lower $^{14}C/C$ ratio value when the animal died supports a fossil age of about 4500 years (the amount of time that has passed since the Flood) while the uniformitarianism approach dates the fossil nearly eight times older (34,380 years). Thus the age discrepancy between mainstream science and the new science explained above is visually clarified. This breakthrough science applies to both AMS and pre-AMS carbon dating results.

A paper[19] written in 2003 by Dr. John Baumgardner and three collaborators contained the following intriguing hypothesis about the unreliability of radioactive geologic dating methods (highlights are the author's, not Baumgardner's)—

"***Our hypothesis is that the source of the discrepancy is the interpretational framework that underlies these methods. Could the proposition, articulated 180 years ago by Charles Lyell, that "the present is the key to the past" be suspect?*** *... we are convinced that there is abundant observational evidence in the geological record that the earth has experienced a global tectonic catastrophe of immense magnitude that is responsible for most of the Phanerozoic geological record.* ***We are persuaded it is impossible any longer to claim that geological processes and rates observable today can account for the majority of the Phanerozoic sedimentary record.***"

More New Science that Challenges the Geologic Time Scale

In addition to the new science involving radioactive dating presented above that challenges the geologic time scale shown in Figure 9, three additional challenges are included here. There are others which are not covered in this book. As a matter of fact, new challenges to the geologic time scale seem to be surfacing on a somewhat regular basis as more and more scientists are willing to explore alternatives to the cherished geologic clock of Darwinian evolution.

One challenge arises from the discovery of polonium radiohalos in granite biotite mica by Robert Gentry, a nuclear physicist.[20]

The granite that Gentry studied normally resides below the many layers of sedimentary rock that form the surface of the earth, and is radioactively dated at the Precambrian super eon (i.e., at least half a billion years old). Gentry has studied more than 100,000 *radiohalos* (or radioactive halos) in granite mica samples that were taken from various locations around the world and from all levels of the geologic clock.

As radioactive uranium atoms in granite decay into lead, they undergo a series of decay steps that give off different levels of energy—which is emitted in all directions. This energy damages the crystal structure of the mica that surrounds a cluster of decaying atoms, producing a series of concentric spheres. The damage can be seen with a microscope as a halo of concentric circles on a thin, planar sheet of mica. Each element in the decay chain has its own characteristic halo. For example, uranium halos show eight concentric rings while polonium halos have only three (typically). Uranium takes an extremely long time to decay—measured in billions of years. Thus it is plausible that uranium radiohalos could form in granite taking millions of years to solidify from its original molten state (which is the standard explanation given by mainstream science).

While studying radiohalos, Gentry discovered numerous isolated halos that were created exclusively by polonium, a highly unstable radioactive element that exists only briefly in the decay chain of uranium. Polonium was discovered by Pierre and Marie Curie. It decays in a matter of days—many orders of magnitude faster than uranium itself. Therefore, a complete polonium radiohalo will only take 1,380 days to form, but the granite in which it resides must have solidified first.[21]

Since polonium only exists for a very short time (because it is so unstable), how could a polonium halo exist by itself inside a piece of granite that was supposed to have taken millions of years to solidify? If the halo formed before the sample cooled, it would have disappeared in the molten granite. Thus the mica sample was obviously solidified when the polonium halo formed, **but since the polonium was deep inside the granite and it only lasts for a very short time, the granite must have also solidified in a very short time—only days, not millions of years!**

No wonder the 1981 assistant Chief Geologist of the U. S. Geological Survey, Dr. Dalrymple, described[21] polonium radiohalos as *"a very tiny mystery!"*

Of course mainstream scientists scoff at Gentry and his findings, even though they have been published in several highly respected, peer reviewed scientific journals. The scientific evidence given by these scientists to combat Gentry's work (relocation of the polonium atoms by fluid) does not provide a satisfactory explanation of the phenomenon.[21] In 1986, Gentry published a book[22] titled *Creation's Tiny Mystery* that included a challenge for these scientists—

"The experiment being proposed is quite straight forward. The basic chemical elements of a granite, which are well-known, are to be melted, and then allowed to cool to form a synthetic rock. If my colleagues do this experiment so the synthetic rock reproduces the mineral composition and crystal structure of granite, then they will have duplicated or synthesized a piece of granite. By doing this they would have confirmed a major prediction of the evolutionary scenario—they would have demonstrated that granites can form from a liquid melt in accord with known physical laws. I will accept such results as falsifying my view that the Precambrian granites are the primordial Genesis rocks of our planet. Furthermore, if they were successful in producing a single ^{218}*Po halo in that piece of synthesized granite, I would accept that as falsifying my view that the polonium halos in granites are God's fingerprints."*

The challenge has yet to be successfully met.[23]

A second challenge to the geologic time scale involves the decay of the earth's magnetic field.[24, 25, 26]

According to modern science, the earth functions like an enormous electromagnet that creates a magnetic field which extends thousands of miles from the earth's surface in every direction. The source of this magnetic field is thought to be a powerful electric current flowing in the earth's outer molten core (composed primarily of a nickel iron alloy) that is rotating and convecting (transferring mass and heat through swirling fluid currents within the molten alloy). The concept behind this hypothesis is similar to wrapping an iron nail with many turns of insulated wire and hooking the wire to a DC (direct current) power source—when the power is turned on, the nail becomes a magnet, and the strength of its magnetic field is determined by the strength (amperage) of the power source and the number of times the wire is wrapped around the nail. In order for the earth to produce such a strong magnetic field, the amperage of the electric current flowing within its core must be billions of amperes!

Since 1835, scientists have measured the strength of the earth's magnetic field on a global scale. From 1970 on, the measurements have been accurate enough to estimate the total amount of energy that is contained within the field. Measurements show that the field's strength is presently declining with a half-life of 1400 years—which means that the field strength is reduced to half of its former strength every 1400 years. As was discussed in Chapter 1, *everything* in mortality naturally degenerates—including the earth's magnetic field.

A decaying magnetic field also means that the field was twice as strong 1400 years ago, and that its strength doubled each 1400 year period going back in time. The strength of the earth's magnetic field is determined by the total amount of energy that is contained within the earth's core, hence the total energy within the earth's core must also have increased going back in time.

What bearing does this have on the geologic scale?

Calculations show that the electrical heating in the core of the earth would be more than a million times greater than it is now just 30,000 years ago. ***This amount of heat would have destroyed the planet!***

So how could the mortal earth be millions or billions of years old? Christians understand that the mortal earth is about 6,000 years old.

Critics of this conclusion correctly point out that the earth's magnetic field has not always decayed so consistently, so how can anyone determine that today's measured decrease in field strength is truly from decay rather than from another fluctuation or even a reversal? To answer these critics, Dr. Russell Humphreys has written two papers[27, 28] which are beyond the scope of this book.

However, to summarize them at a high level, the measured magnetism found in ancient artifacts such as pottery shows that the present decay rate began about AD 500. For several thousand years before then, the strength of the field fluctuated up and down extensively, as is told by the magnetism of geologic strata (*paleomagnetic* data). During the time that the fossil layers of sediment were being laid down on the earth's surface, which will be shown later to be during the year-long Flood, the earth's magnetic field actually reversed itself many times—about once a week. When all the fountains of the great deep were broken up at the beginning of the Flood, there were strong convective (swirling) motions within the earth's core that released significant amounts of energy, causing the rapid reversals of the earth's magnetic field. (Note: such magnetic field reversal cycles have been observed in the sun.)

After the Flood, the residual turbulent flow of the earth's molten core created havoc with the earth's magnetic field, which resulted in more than two poles (i.e., more than only one north and one south pole—this is confirmed by paleomagnetic data). Over time, the turbulence subsided and the earth returned to its dipole (i.e., one north and one south pole) magnetic field. But while the turbulence was subsiding, fluctuations of field strength occurred as diminishing magnetic polarities opposed each other. The earth's magnetic field strength reached its maximum about the time of Christ, and has been decaying ever since.

The essence of Dr. Humphreys' papers is that the earth's magnetic field has always been dissipating energy (i.e., obeying the Second Law of Thermodynamics), despite the fluctuations and reversals described above, ***and therefore it cannot be more than 9,000 years old.***

As a side note, Dr. Humphreys also developed a model during the 1980s for the creation of magnetic fields in planets throughout the Solar System. This model assumes that the cores of all the planets, including earth, were initially composed of water with the hydrogen nuclei (single protons) all spinning in the same direction—which created the planets' magnetic fields. Remarkably, his model correctly predicted the strength of magnetic fields coming from Uranus and Neptune. The model also correctly predicted the decay rate of Mercury's declining magnetic field. Measurements of the strength of Mercury's magnetic field were taken in 1974 by the Mariner 10 spacecraft and again in 2008 by the Messenger spacecraft. The measurements showed a decline of 4% in the strength of Mercury's magnetic field, which is precisely what Humphreys' model predicted.[29, 30]

Water within the earth

It now appears that Dr. Humphreys' model for the creation of magnetic fields in planets was actually a scientific breakthrough. New science involving the interior of the earth indicates that water is found *everywhere* inside the earth. Since water is composed of two of the three most abundant elements in the universe (hydrogen and oxygen), the pervasive presence of water within the earth should be no surprise. Furthermore, the scriptures in Genesis that describe the creation of the earth refer to "the waters" multiple times, implying the unique significance of water in the process of creation.

A groundbreaking paper[71] titled "Oceans of Water in the Earth's Core" by S. K. Saxena et al was presented at the American Geophysical Union's Spring Meeting in 2002. This paper stated, *"...it is essential that we consider water as an important core substance for the reasons that it is abundant and suitably light to decrease the density and melting temperature of iron. Furthermore, if water is indeed stored in the core, it should provide a suitable source for water in the mantle and affect the dynamics of melting everywhere."* The paper went on to describe experiments that had been conducted to understand iron-water chemistry at the high pressures and temperatures thought to exist within the earth's core. The tests confirmed that water reduced the melting temperature of iron by approximately 1000° Kelvin (K). ***The paper concluded that even if only one percent of the earth's core is water, that will amount to ten times more water than is presently found in all the oceans.***

A second article[75] titled "Huge 'Ocean' Discovered Inside Earth" was posted online in 2007 at livescience. This article opened with, *"Scientists scanning the deep interior of Earth have found evidence of a vast water reservoir beneath eastern Asia that is at least the volume of the Arctic Ocean. The discovery marks the first time such a large body of water has been found in the planet's deep mantle."* The discovery was made by Michael Wysession, a seismologist at Washington University in St. Louis, and his former graduate student, Jesse Lawrence, after analyzing more than 600,000 seismograms (records of waves generated by earthquakes traveling through the earth). According to Wysession, the composition of some ocean floor rocks is up to 15% water. If these rocks sink into the earth's mantle (more about this will be covered later in this chapter), *"Water inside the rock goes down with the sinking slab and it's quite cold, but it heats up the deeper it goes, and the rock eventually becomes unstable and loses its water."* The water then rises into the overlying mantle region.

A third groundbreaking paper[72] that also dealt with water within the earth's mantle was published in the journal *Nature* in March of 2014. This fascinating paper opened with, *"The ultimate origin of water in the Earth's hydrosphere is in the deep Earth ..."* The paper discussed the first-ever terrestrial sample of ringwoodite that had been taken from Brazil in 2009. The sample was part of a volcanic rock (kimberlite) that is known to come from the deepest level of volcanic rocks within the earth's mantle. Analysis of the mineral confirmed that 1.5% of the sample's weight was due to water that had been absorbed by the ringwoodite (probably in a molecular form called hydroxyl). The sample came from the earth's transition zone, which is midway between the upper and lower boundaries of the earth's mantle (250 – 400 miles/400 – 660 km deep), and provides *"irrefutable proof that the mineral came from a deep water reservoir."*[73] An article[74] posted online March 12, 2014 by the University of Alberta (where the scientist who identified the sample works) added the following enlightening information—*"Knowing water exists beneath the crust has implications for the study of volcanism and plate tectonics, affecting how rock melts, cools and shifts below the crust."* The lead scientist on the international team that analyzed the sample, Graham Pearson, commented, ***"One of the reasons the Earth is such a dynamic planet is the presence of some water in its interior. Water changes everything about the way a planet works."***

In June 2014, another paper[77] titled "Dehydration Melting at the Top of the Lower Mantle" was published in the journal *Science.* A National Public Radio online article[76] reviewing this paper suggested that this water in the earth's mantle could be the original source of water in the oceans and made this claim: ***"If just 1 percent of the weight of mantle rock located in the transition zone was water it would be equivalent to nearly three times the amount of water in our oceans."***

These three discoveries (not an exhaustive list) provide new understanding about the vast amount of water that exists within the interior of the earth. Over time, a greater understanding of all this subsurface water and its physical ramifications will undoubtedly open new doors to science. One such opened door, regarding catastrophic plate tectonics, will be discussed below.

A third challenge to the geologic time scale involves continental erosion rates.

In 2011, a paper [32] was published about the rate at which the earth's surface is eroding. This paper was written by Eric Portenga and Paul Bierman, recipient of the Donath Medal ("the most promising young geologist in the United States"). Both authors are from the University of Vermont. They compiled, normalized, and compared nearly 1600 measurements of *in situ* (meaning "in its natural place of origin") radioactive beryllium in order to calculate how fast the earth's surface is eroding. They found that the average erosion rate of outcrops is about 40 feet per million years. For reference, the average thickness worldwide of continental crust above sea level has been estimated at about 2050 feet.[33] ***At an average erosion rate of 40 feet per million years, the mortal earth's continents would have eroded to sea level in about 51.25 million years. According to the geological time frame, the continents came into existence about 3.5 billion years ago; therefore, they could have eroded to sea level 68 different times since they were supposedly formed!***

Portenga and Bierman also found that drainage basins erode even faster—an average of 717 feet per million years. However, they only represent about 2.3% of the world's land area. More importantly, Portenga and Bierman also wrote that—

"Paleo-erosion rates should be higher than modern rates in warmer climate locations that cooled significantly during the Pleistocene."

The Pleistocene is the geological *epoch* that lasted from about 2.588 billion to 11,700 years ago. ***This means that an average erosion rate of 40 feet per million years is too conservative. For the first billion years or so that the continents supposedly existed, they eroded even faster!***

According[33] to Dr. Ariel A. Roth—

"It has been suggested that mountains still exist because they are constantly being renewed by uplift from below. However, this process of uplift could not go through even one complete cycle of erosion and uplift without eradicating the layers of the geologic column found in them. Present erosion rates would tend to rapidly eradicate evidence of older sediments; yet these sediments are still very well represented, both in mountains and elsewhere."

Dr. Roth has also written that about 31 billion tons of sediment are carried by rivers into the world's oceans every year. The oceans have a total volume of about 372 million cubic miles, ***which would be completely filled with sediment in just 114 million years at this rate of erosion!***

Thus erosion rates of the earth's land surfaces do not support the hypothesis that the mortal earth is hundreds of millions or billions of years old.

Discussion of New Science and the Geologic Time Scale

The following significant points have been presented so far in this chapter—

1. As with Darwinian evolution, mainstream science is committed to uniformitarianism and the geologic time scale shown in Figure 9.
2. The measured rate of helium diffusion in zircon crystals from deep underground granite was compatible with the crystals being about 5,000 years old—yet uranium-lead dating of the same granite resulted in an age of 1.5 billion years.
3. There are more than a dozen known locations where historical volcano lava flows were dated millions of years older than their actual ages, proving that dating methods involving argon are not reliable.
4. Using the high precision accelerator mass spectrometer (AMS) carbon dating method during the last twenty seven years, no fossils have been found anywhere in the world that are dated as old as 90,000 years. Rather, most fossils seem to have approximately the same $^{14}C/C$ ratio, which indicates an age of roughly 50,000 years. However, 50,000 years is too old because the rate of ^{14}C production in the atmosphere actually exceeds the rate of ^{14}C decay by 30%. Using other radioactive dating methods, these same fossils are dated up to 542 million years.
5. AMS carbon dating results imply that all fossilized organisms lived on the earth at the same time.
6. The recent discovery of polonium radiohalos deep inside samples of granite indicate the granite must have solidified in a very short time—only days, not millions or billions of years.
7. Recent studies of the earth's decaying magnetic field indicate that the mortal earth cannot be more than 9,000 years old.
8. Erosion rates of the earth's land surfaces do not support the hypothesis that the mortal earth is hundreds of millions or billions of years old.

Taken together, perhaps the most important truth to emerge from these points is that science cannot reliably date the age of the earth, nor the geologic time scale, nor the fossil record.

Different dating methods produce different results. Recent volcano lava flows do not date properly. Modern fossil dating does not agree with the geologic time scale. Dating method assumptions cannot definitely explain discrepancies. The decaying magnetic field of the earth, along with earth's present erosion rates, do not support dating the mortal earth at millions or billions of years.

As a result of this rather muddled reality, a scientist can make a case to support whatever agenda he or she personally wishes to promote. Thus, evolution scientists point to scientific dating results that seem to support evolution timescales, while creation scientists point to scientific dating results that seem to support a young earth only thousands of years old. *However, both cannot be right.*

To shine more light on this muddled reality, consider the prophetic words that the Apostle Peter wrote, which provide the needed guidance for resolving this frustrating dilemma of the last days (highlights are the author's)—

Knowing this first, that there shall come in the last days scoffers, walking after their own lusts, And saying, Where is the promise of his coming? for since the fathers fell asleep, ***all things continue as they were from the beginning of the creation. For this they willingly are ignorant of, that by the word of God the heavens were of old, and the earth standing out of the water and in the water: Whereby the world that then was, being overflowed with water, perished: But the heavens and the earth, which are now,*** *by the same word are kept in store, reserved unto fire against the day of judgment and perdition of ungodly men. But, beloved, be not ignorant of this one thing,* *that one day is with the Lord as a thousand years, and a thousand years as one day.* (2 Peter 3:3-8)

Note that Peter portrays those who believe in uniformitarianism as being *"willingly ignorant of"* the effects of the Great Flood, which caused the antediluvian world to perish. Peter also alludes to *"the heavens, and the earth which are now,"* drawing a clear distinction and boundary between the two worlds. His words imply an earth before the Flood that was vastly different from the earth on which we live today.

Thus the Flood holds the key to resolving the dilemma discussed above.

Earth's Catastrophic Past

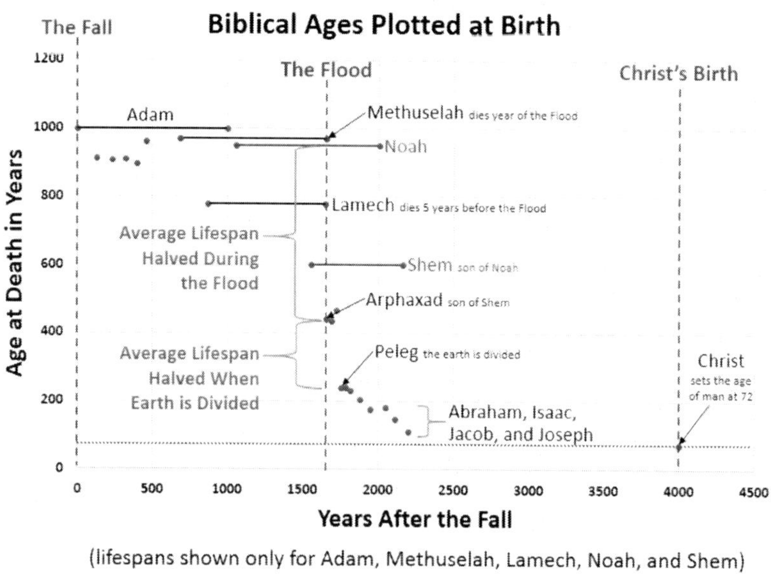

Figure 11—Biblical ages showing declining lifespans with two different catastrophic reductions caused by the Flood and the Division of the earth. Average lifespan was reduced by half as a result of these catastrophic events.

Chapter 2 provided scientific proof that Darwinian evolution does not occur, and that the human genome is decaying. Figure 8 (in Chapter 2) illustrates the decline in fitness and associated lifespan that mankind has experienced since the Flood.

Figure 11 on the previous page also illustrates declining lifespans of mankind, but includes more of the scriptural record, and begins with the Fall of Adam and Eve instead of the Flood. Lifespans of Adam, Methuselah (the grandfather of Noah), Lamech (the father of Noah), Noah, and Shem (a son of Noah) are shown in order to provide a more visual representation of mankind's history.

Eight of the first nine patriarchs included in Figure 11 lived at least 895 years. It is not known if Lamech died from old age or not; regardless, the maximum lifespan he could have had is only 782 years because he would have died in the flood (unless taken aboard Noah's ark); therefore his lifespan should not be averaged when plotting the true decay of human lifespan. Removing Lamech from the group of patriarchs up to and including Noah, *the average lifespan immediately after the Fall was about 938 years, and it persisted for more than 1600 years.*

According to Genesis, Shem was 100 when Arphaxad was born, and this was two years after the flood, placing Shem's birth at 1,559 years after the Fall. Thus Shem was relatively young (97) when the Flood occurred, compared to the average lifespan in effect when he was born (938 years).

More importantly, Shem only lived 600 years—64% of the typical lifespan that prevailed when he was born. Shem's son Arphaxad, who was born two years after the flood, lived a total of 438 years, and the next two patriarchs lived 433 and 464 years. Thus *the average lifespan for the first three generations of Shem's posterity was 445 years*—***only 47% of the average lifespan that prevailed before the Flood!***

Noah and Shem were the only patriarchs included in Figure 11 who lived through the Flood. Shem's lifespan appears to be a transition between the lifespan of his father Noah (950 years) and that of his posterity for three generations (445 years averaged). Notice that Shem was relatively young when he experienced the Flood, while Noah was already past the midpoint of his life.

After 1,657 years of mankind's lifespan averaging 938 years, the lifespan of men who were born after the Flood suddenly dropped to an average of 445 years—a 53% reduction; and this incredible discontinuity occurred in a mere one year time frame! Clearly, something catastrophic must have happened to the earth in order to bring about such an abrupt and remarkable change in the typical lifespan of mankind.

But this is not the only catastrophic period in the earth's history that is revealed in Figure 11. Notice that Peleg only lived 239 years. The book of Genesis records the following about Peleg—

> *... for in his days was the earth divided ...* (Genesis 10:25)

In Hebrew, Peleg means "division." Peleg was the son of Eber, who lived to be 464 years of age. During the three generations that began with Peleg, the patriarchs lived 239, 239, and 230 years. Thus *the average lifespan for the first three generations of Eber's posterity was 236 years*—***only 53% of the average lifespan that prevailed before the earth was divided!***

In summary, the Flood, although it only lasted for about a year, cut the average lifespan of mankind in half, and this happened 1657 years after the Fall. Likewise, the Division of the earth that occurred in the days of Peleg cut the average lifespan in half once again—approximately a century after the Flood. Clearly, something catastrophic must have been going on with the earth during these two historical events in order to cause such drastic reductions in the average lifespan of mankind.

After the Division of the earth occurred, no such drastic reductions in lifespan have been recorded in history or scripture. Since then, the average lifespan has been fairly flat, as shown and predicted in Figure 8. At times, however, social conditions have either lowered it (e.g., during the dark ages when health conditions were deplorable) or raised it (as medical technology is currently doing). Nevertheless, the human genome continues to degenerate.

In order to reduce mankind's lifespan by half in only a year (during the Flood), there must have been highly accelerated damage to the human genome that took place that year. Furthermore, the damage must have become genetically fixed (made permanent) within a single generation, consisting of the three sons of Noah and their wives. As explained in Chapter 2, some forms of radiation and certain chemicals are known to cause genetic mutations, so one or both of these agents may have been involved. Most likely, global cosmic radiation was the agent, as will be discussed in more detail below.

Before the Division of the earth, all land masses were connected in a single supercontinent. Mainstream science has given the name "Pangaea" (in Greek, "entire Mother Earth") to this supercontinent,[36] and has even provided a map of how it may have looked, as shown here. Pangaea supposedly formed about 300 million years ago, during the late Paleozoic era, and started to break up about 200 million years ago, in the early Mesozoic era.

The Supercontinent Pangaea
(courtesy en:User:Kieff, Wikimedia Commons)

Scientific evidence that Pangea existed is provided in the fossil record, the geologic record, and the continuity of mountain chains. For example, fossils of a small animal called *Lystrosaurus* have been found in Africa, India, and Australia. Also, fossils of a freshwater reptile dubbed *Mesosaurus* have only been found in the coasts of Brazil and West Africa. Likewise, matching geological trends have been discovered on the eastern coasts of South America and the western coasts of Africa. Furthermore, the Appalachian Mountain chain extends from the southeastern United States to the Caledonides of Ireland, Britain, Greenland, and Scandinavia.

The Division of the earth involved an actual separation of the continental land masses. In addition, many of the mountain chains that exist today may have been created by tectonic upsurges during this catastrophic period of time. For example, the pinnacle of Mount Everest, currently the tallest mountain on earth (standing at 29,029 feet (8,848 m) above sea level), was once at the bottom of the ocean, as revealed by the fossils of trilobites and other ancient sea life near its limestone summit.

Mount Everest
(courtesy Rdevany, Wikimedia Commons)

Many historians and creation scientists believe the Division of the earth associated with Peleg involved only people rather than land masses, or that it involved major rivers. They also believe that today's continents were formed during the Flood. However, such a position does not explain the catastrophic event that took place in the days of Peleg as revealed in Figure 11. As with the Flood, there must have been highly accelerated damage to the human genome taking place during the Division of the earth, and the damage must have become genetically fixed (made permanent) within a single generation.

The Creation Research Society

As pointed out in Chapter 1, a significant area of disagreement between the scriptures and mainstream contemporary science is the extent of the Flood God sent to cleanse the earth. The scientific community does not believe that the Flood was global; rather, it is thought to have been nothing more than a regional flood in the Mesopotamian Valley—if it occurred at all. Most scientists believe the Bible's account of the Flood is a myth.

However, not all scientists agree with mainstream science regarding the Flood. As Dr. John Baumgardner and his three collaborators hypothesized in the carbon dating section above, *"...we are convinced that there is abundant observational evidence in the geological record that the earth has experienced a global tectonic catastrophe of immense magnitude that is responsible for most of the Phanerozoic geological record."* Of course, they are referring to the Great Flood of Noah's day.

Dr. Baumgardner and his colleagues are members of the Creation Research Society—a professional organization of trained scientists who are committed to "scientific special creation." The Society was organized in 1963 and has grown into an organization with worldwide membership. All members must subscribe to the CRS Statement of Belief—

1. The *Bible* is the written Word of God, and because it is inspired throughout, all its assertions are historically and scientifically true in the original autographs. To the student of nature this means that the account of origins in *Genesis* is a factual presentation of simple historical truths.

2. All basic types of living things, including man, were made by direct creative acts of God during the Creation Week described in *Genesis*. Whatever biological changes have occurred since Creation Week have accomplished only changes within the original created kinds.

3. The great flood described in *Genesis*, commonly referred to as the Noachian Flood, was an historic event worldwide in its extent and effect.

4. We are an organization of Christian men and women of science who accept Jesus Christ as our Lord and Savior. The account of the special creation of Adam and Eve as one man and one woman and their subsequent fall into sin is the basis for our belief in the necessity of a Savior for all mankind. Therefore, salvation can come only through accepting Jesus Christ as our Savior.

Considering this Statement of Belief, it should be clear that scientific work performed by the Creation Research Society regarding the Flood should be welcomed with joy and open arms by all Christians—as new and breakthrough science that expands the eternal circle of truth.

Scientific work regarding the Flood that has been performed by members of the Creation Research Society was succinctly and admirably abridged by Dr. John Morris in the book *The Global Flood—Unlocking Earth's Geologic History*.[36] The relevant, salient points of this book are reviewed in this chapter. The author of this book recommends *The Global Flood* to readers who desire more detail than can be provided here.

Since no scientist was available to observe what actually happened during the Flood, only hypotheses based on field observations can be presented. However, the hypotheses reviewed here are not biased by an ideological commitment to both forward Darwinian evolution and uniformitarianism, as are the hypotheses of mainstream science. Admittedly, these hypotheses are founded on the beliefs that the Flood as described in Genesis actually occurred and that it was global in extent. Fortunately, the Flood left extensive evidence of its catastrophic work that is readily observable, so there are actual scientific breakthroughs behind the hypotheses presented below.

Most importantly, as these hypotheses are discussed, the reader should ask himself/herself which hypotheses make the most sense—those developed by mainstream science or those presented here.

Earth Conditions Before the Flood

Before discussing the Flood and its effects on the earth, it will be enlightening to first review what science has learned about pre-Flood conditions. Understanding what the earth was like before it was destroyed by the Flood will enable the reader to better understand how the Flood could have actually done what is recorded in the book of Genesis.

The earth's pre-Flood magnetic field

As already discussed above, the strength of the earth's magnetic field has been decreasing ever since the Fall of Adam and Eve. This means that the strength of the field was at its highest level immediately after the Fall. The Flood began 1,656 years after the fall, which is only 18% longer than the currently measured half-life (decay rate) of the earth's magnetic field (remember that field strength is reduced by half every 1,400 years). Since the Flood ended, approximately three more half-lives have passed, thus a total of roughly 4.3 half-lives have occurred since the Fall. With this information, one can calculate that the strength of the earth's magnetic field is presently only 5.4% as strong as it was after the Fall (assuming the decay rate has not changed significantly). This means that the field was about 18.5 times stronger right after the Fall than it is today! At the beginning of the Flood, it would have been approximately 8.4 times stronger.

A magnetic field 8.4 to 18.5 times stronger than today's field would have prevented much more cosmic radiation from penetrating the atmosphere and causing genetic damage to all living things.

Thus the human genome would have experienced far fewer mutations per generation than it currently encounters—which undoubtedly was a contributing factor to the very long lifespans that mankind enjoyed before the Flood.

The earth's pre-Flood climate

The fossil record includes many trees that were alive before the Flood. Nearly all layers of today's sedimentary rock were deposited as sediments during the Flood, so living trees were buried and well preserved by those sediments. Over time, these trees petrified. Studying their trunk rings has shown that nearly every year before the Flood was a very good growth year, providing the tree with abundant nutrition and water.[38] In fact, *"A somewhat monotonous regularity testifies to an apparent subtropical 'paradise' on all continents and in each 'time' period."* Further, *"Until the Flood, planet earth's physical systems enjoyed a period of relative equilibrium, for the biblical record contains no mention of earthquakes, volcanic eruptions, or natural disasters."* [38]

Evidently, the earth's pre-Flood climate was delightfully subtropical over all its land masses, with temperatures that always stayed above freezing (32°F or 0°C).

The only places on earth today that have a subtropical climate are shown here (lighter shade).

Areas of modern earth with a subtropical climate
(courtesy Maphobbyist, Wikimedia Commons)

The earth's pre-Flood topography

One area of science where uniformitarian scientists and creation scientists agree (up to a point!) is the relative timing of when today's mountain ranges were formed. Geologists have determined that they arose very late in earth's history. From the Swiss Alps to the New Zealand Mountains, a total of twenty-nine mountain ranges have been characterized as having their main uplift occur within the last five million years; some even as recently as one million years.[41] In the geological time frame, five million years is very young! Creation scientists place the uplift timing of these same mountain ranges at the end of the Flood or shortly thereafter. As discussed above, this likely happened during the Division of the earth (about a century after the Flood).

Based on this information, it is reasonable to assume that the mountains which existed before the Flood were not as high as today's mountains. Furthermore, the ocean bottoms were not as deep as they are today. *"There is also evidence that marine plants and animals, whose fossils are found in great abundance all around the earth, lived near the coast in relatively shallow seas, not under great depths of water."*[40]

Thus the topography of the pre-Flood earth was considerably flatter than it is now.

Geologists have calculated that, if the earth's crust, including land masses and ocean bottoms, were all at the same radius from the center of the earth (forming a smooth sphere), the amount of water and ice currently found in the oceans and on land would cover this smooth sphere to a depth of approximately 10,000 feet (3 km).[41] This figure provides a useful reference point when contemplating the global extent of the Flood.

Breakthrough Science Regarding Plate Tectonics

In 1858, the French scientist and geographer Antonio Snider-Pellegrini published a book titled *The Creation and Its Mysteries Unveiled*. This book contained the illustrations shown below. They depict his theory of how the continents may have formed a supercontinent before they were separated. (Note: "avant la séparation" means "before the separation" and "aprés la séparation" means "after the separation.") Snider-Pellegrini developed this theory after finding identical plant fossils in Europe and the United States. He also found other matching fossils on all the continents. Snider-Pellegrini believed that the Flood caused the continents to separate.[42] As it turns out, he was way ahead of his time. For example, he was 54 years ahead of Alfred Wegener's theory of continental drift that was adopted by mainstream science in the 1950s. Snider-Pellegrini also had great foresight attributing the Division of the earth to the Flood. Ironically, his book was published in the same year that Darwin published *On the Origin of Species by Means of Natural Selection*.

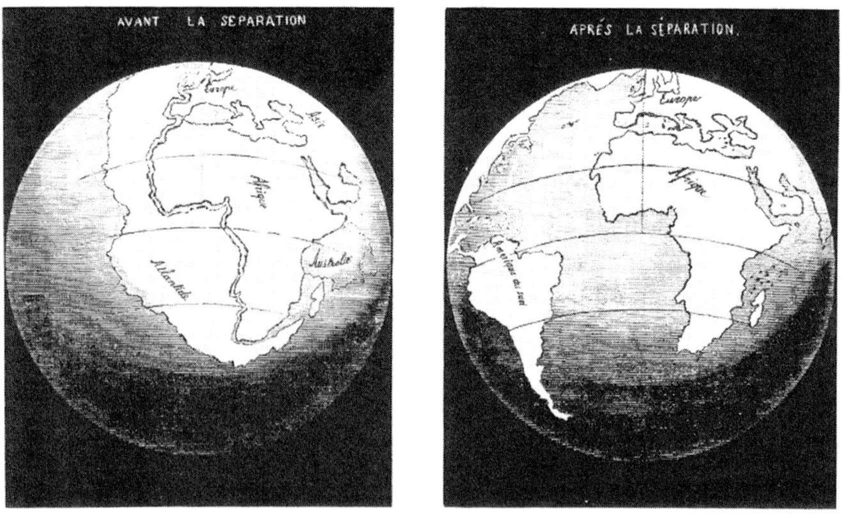

Illustrations from Antonio Snider-Pellegrini's book
The Creation and Its Mysteries Unveiled.
courtesy Wikimedia Commons—published before 1923 and public domain in the U.S.

Today, mainstream science attributes continental movement to the theory of *plate tectonics,* which was developed largely during the 1960s. This theory describes large scale motions of the earth's *lithosphere,* which comprises the earth's *crust* and the uppermost, solid layer of the earth's *mantle* (see Figure 12 below). Note that the lithosphere is thicker for land surfaces than for oceans.

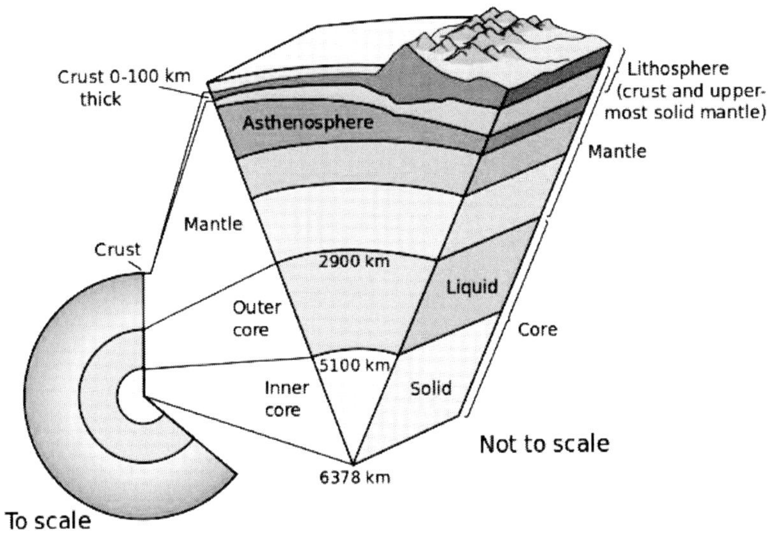

Figure 12—Earth cutaway from core to crust, showing the lithosphere which comprises the crust and the uppermost mantle layer.
courtesy U.S. Geological Survey, Wikimedia Commons (in the public domain)

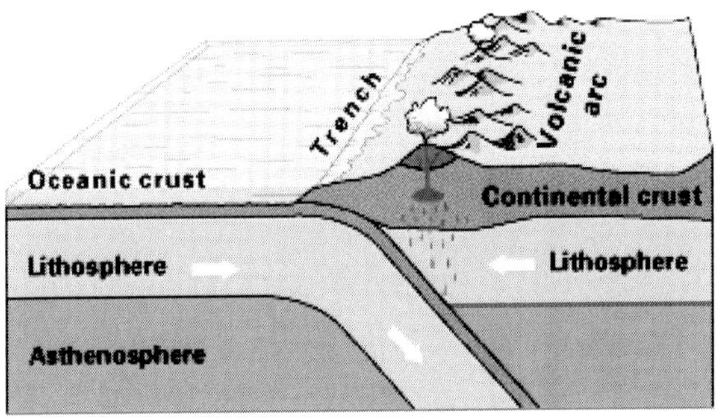

Figure 13—Subduction zone
courtesy U.S. Geological Survey
(in the public domain)

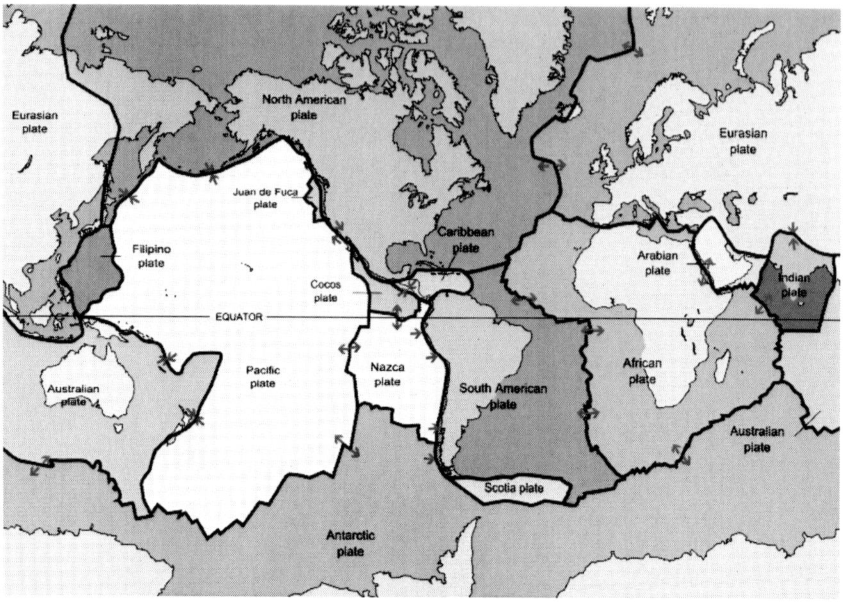

Figure 14—Tectonic plates that form Earth's lithosphere.
courtesy U.S. Geological Survey, Wikimedia Commons (in the public domain)

According to the theory of plate tectonics, the lithosphere is made of major and minor *tectonic plates* (see Figure 14 above). Where these plates meet, their motion relative to each other determines what type of boundary is formed.

A convergent (or destructive) boundary is formed when plates move toward each other and collide. Either a continental collision occurs or a *subduction zone* is created. In a subduction zone, the subducting plate (normally with oceanic crust) moves beneath the other plate (see Figure 13 on the previous page). As a result of the friction, pressure, and heat caused by subducting plate material melting within the earth's mantle, volcanos and earthquakes are common near convergent boundaries. If a continental collision occurs (one continental plate colliding with another continental plate), large mountain ranges are formed inland (such as the Himalayas).

If the tectonic plates are moving away from each other, a divergent (or constructive) boundary is formed. The divergence creates *rifts* in land surfaces and *ridges* in ocean bottoms. If a divergent boundary produces gaps which are then filled with rising molten lava, volcanic islands are formed. The molten lava comes from convection (swirling mass transfer) within the molten part of the earth's mantle.

According to plate tectonics theory, the amount of lithospheric material that is lost through subduction at convergent boundaries is balanced by the formation of new oceanic crust via *seafloor spreading* along divergent boundaries. In this way, the total surface area of the earth remains constant.

The plates move because the lithosphere is stronger than the *asthenosphere* portion of the earth's mantle that lies directly below it (see Figures 12 and 13). Plate movement is driven by a combination of (*i*) differences in gravitational forces that result from movement of the sea floor away from spreading ridges at divergent boundaries (which in turn are caused by variations in the topography and density of the oceanic crust); and (*ii*) downward suction that occurs at subduction zones. Presently, relative lateral (horizontal) movement of the tectonic plates varies from zero to four inches (10 cm) yearly.

Of course, modern science's doctrine of uniformitarianism demands that no catastrophic events have been involved with plate tectonics, hence the continents have been drifting apart for roughly 200 million years.[43]

A groundbreaking paper [44] published in 1994 by six members of the Creation Science Fellowship, Inc. provided a bridge between the theories of Antonio Snider-Pellegrini and modern plate tectonics. Titled *Catastrophic Plate Tectonics: A Global Flood Model of Earth History,* this paper pulled together the work of many scientists to develop a state-of-the-art model based on rapid plate tectonics that explains geological evidence which had not been satisfactorily explained by conventional plate tectonics theory, nor by any other earth history model.

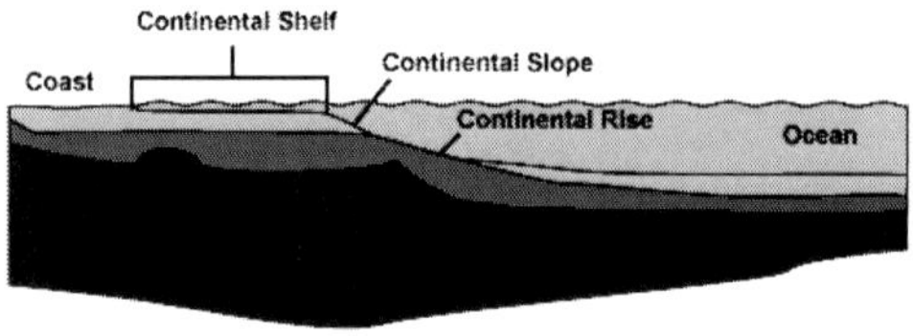

The continental shelf, slope, and rise comprise the continental margin
courtesy Wikimedia Commons (in the public domain)

According to this new model, the Flood began when slabs of oceanic lithosphere broke loose and subducted (slid underneath the continental lithosphere) along the thousands of miles of pre-Flood *continental margin*. As shown here, the continental margin is a zone of the ocean floor that includes the continental shelf, the continental slope, and the continental rise.[45] Remember that the pre-Flood earth had only a single supercontinent.

When the slabs subducted beneath the supercontinent's lithosphere, deformation of the upper boundary of the earth's mantle (see Figure 12) raised the temperature and lowered the *viscosity* (i.e., liquid "thickness"—honey has a higher viscosity than water) of the mantle surrounding the penetrating slabs. The lower viscosity then allowed an increase in subduction speed, which in turn heated up the surrounding mantle even more. This escalating process eventually resulted in a *thermal runaway instability* that enabled yards-per-second (meters-per-second) subduction of the ocean's lithosphere. Such runaway subduction bent the ocean plate beneath the supercontinent plate, producing a trench along the continental margin of the pre-Flood supercontinent that is similar to the ocean trenches found today at the eastern, northern, and western continental margins of the Pacific Ocean (see Figure 14).

Virtually all of the pre-Flood ocean's lithosphere was subducted this way during the Flood, essentially being recycled

back into the earth's mantle. Present age-dating of the earth's ocean floors confirms this fascinating phenomenon.

As the pre-Flood ocean's lithosphere was being subducted in this manner, the supercontinent lithosphere slabs attached to these ocean slabs were also pulled toward the subduction zones, causing the supercontinent to strain and eventually fracture into multiple continents. The continents moved horizontally at yards/meters-per-second speeds. Collisions of these continents at subduction zones created fold-and-thrust-belts, such as the mountainous foothills adjacent to the Appalachians, Himalayas, and Swiss Alps.

The rapid subduction of the ocean's lithosphere initiated large-scale convective flow (swirling and vertical) of molten magma throughout the entire mantle of the earth. Seismic tomography studies have confirmed this phenomenon. (Note: seismic tomography creates 3D images of the earth's internal velocity structure, which varies by chemical composition, thermal structure, and depth in the earth. Earthquakes provide the seismic events required for seismic tomography measurements.)

Note the results of one such seismic tomography study[46] by Adam Dziewonski, Professor Emeritus at the Harvard Seismology Group (highlights are the author's, not Dziewonski's)—

"The most striking large-scale three-dimensional feature of the model is a ring of high velocities circumscribing the Pacific basin within a depth range from 1000 km to the core-mantle boundary. If the proportionality between perturbations in the velocity and temperature is assumed, the derived pattern of anomalies could be consistent with a two boundary layer model of convection in the lower mantle. … ***a significant global increase in the level of perturbations in the lowermost mantle is established beyond question.***" (Note: a

perturbation is a deviation of a system from its normal state, caused by an outside influence.)

Also consider the following from the abstract[47] of another paper written by authors from Western Washington University and the University of California, Berkeley (highlights are the author's)—

"A compilation of lithosphere consumed globally in the past 180 m.y. [million years] *reveals that* ***the equivalent of the surface area of the Earth has descended into the mantle****, producing "lithospheric graveyards."*

As the pre-Flood ocean's lithosphere was transported to the earth's mantle/core boundary, its cooler temperature cooled the earth's molten outer core (see Figure 12), which led to convective flows within it. These swirling flows produced rapid geomagnetic reversals in the outer core, which in turn reversed the earth's magnetic field at the earth's surface. Water within the core undoubtedly aided this convection.

This phenomenon predicts and explains the magnetic reversals imprinted in lava flows of the U.S. Northwest, as well as other curious discoveries in the intensity and direction of residual magnetization found in ancient rocks and artifacts.

As the pre-Flood ocean's lithosphere subducted, it produced *spreading centers* on the ocean floor that were thousands of miles long. Upwelling, molten mantle material (magma) rose to the ocean floor's surface at these spreading centers, vaporizing the ocean water and producing geysers of superheated gases that extended the entire length of the spreading centers. Jettisoning their gases high into the atmosphere, these spouting geysers may have been the *fountains of the great deep* mentioned in the book of Genesis. The most abundant substance spewing from the geysers was water vapor, which cooled as it shot higher and higher into the atmosphere. When it condensed, the resulting water fell to the earth as pounding rain. This process may have been the *windows of heaven* opening, as described in the book of Genesis, that continued without abatement for the first 150 days of the Flood.

The rapid upwelling of magma actually raised the level of the ocean floor along these spreading centers, producing the world's longest mountain chain that is now called the *mid-ocean ridge*. As illustrated here, a rift (valley) runs along the spine of each mountain range in this underwater mountain system, which is 49,700 miles (79,984 km) in total length. *Black smokers*, such as the one shown in this illustration, are hydrothermal vents that spout hot water saturated with dissolved minerals coming from the earth's mantle. Such smokers may be remnants of the fountains of the great deep.

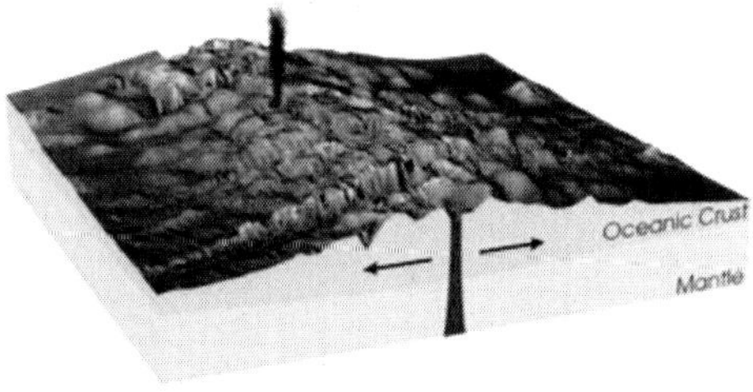

Section of the mid-ocean ridge—a spreading center with a black smoker
courtesy NASA, Wikimedia Commons (in the public domain)

As the level of the ocean floor continued to rise abruptly along these oceanic spreading centers during the Flood, the displaced ocean water was forcefully pushed onto the supercontinent, inundating everything in its path. *The sea level itself probably rose a mile (1.6 km) or more as a result of the upwelling magma.*

Today, most Flood sediments are found on the continental margins and on the continents themselves, rather than on the ocean floor. Several phenomena contributed to the transport of Flood sediments from ocean bottom to the continental margins and farther. First, the subducting ocean lithosphere carried sediments to the subduction zones, which also moved them toward the supercontinent. Second, as the ocean plates were forced to rapidly bend into the earth's mantle at the subduction zones, they warped upward, sloughing off the sediments they were carrying with them. This sloughing process transported the sediments to a higher sea depth, making it easier for the turbulent ocean waters to wash them onto the supercontinent. Third, rapid subduction of the ocean plates also pulled the continental margins downward, making it easier for the ocean waters to wash the sediments onto the continental margins. Fourth, as the relatively cold pre-Flood ocean lithosphere was replaced with hotter, less dense rock coming from the mantle below, the ocean bottom gradually rose. Once again, this made it easier for the churning ocean waters to wash the sediments onto the continental margins and the supercontinent. Fifth, the dynamics of mega-sized waves hitting the continental shelf transported the sediments shoreward and onward into the interior of the supercontinent.

Global volcanic activity associated with such rapid tectonics was of unprecedented magnitude. At the mid-ocean ridge, the relative speed of tectonic plates moving away from each other (yards/meters per second) produced a violent volcanic source thousands of miles in length. The heating and melting of subducted sediments would also produce explosive volcanic eruptions of rock rich in silica and alumina, as is found in the Andes Mountains of South America, the Cascade Mountains of the United States, and the Aleutian, Japanese, Indonesian, and New Zealand islands in the Pacific Ocean. More about the volcanic activity resulting from catastrophic plate tectonics will be discussed below.

The rapid bending of ocean lithosphere—which occurred as the subducting plates shot downward into the earth's mantle—produced frequent, powerful earthquakes at the subduction zones. Earthquakes also resulted from colliding tectonic plates, volcanic activity, and gravitational equilibrium adjustments between the earth's lithosphere and the asthenosphere directly below it. All this earthquake activity provided energy to fracture solid blocks of rock and encourage thrust faulting, which pushes rock from lower strata up and over rock in higher strata.

Once the pre-Flood ocean floor had been replaced with newer, less dense rock, the catastrophic plate motion ended. Since magma no longer upwelled at the oceanic spreading centers, the global rain also stopped. This likely happened 150 days after the Flood began—when the *"fountains also of the deep and the windows of heaven were stopped, and the rain from heaven was restrained."* (Genesis 8:2) It has been calculated that the gravitational potential energy released from the subducting ocean slabs was sufficient to drive the dynamics of the entire Flood.

After all the rapid horizontal motion ceased, cooling increased the density of the new ocean floor, which then gradually dropped and formed the ocean basins that exist today. As the waters receded from the newly formed continents, the most superficial and least solidified sedimentary deposits were eroded off the continents. This wind-assisted erosion (see Genesis 8:1) left a break in the continental sedimentary geologic record (an *unconformity*) that is not reflected in the associated ocean sedimentary geologic record. The absence of these sedimentary deposits on the continents likely explains the dearth of human, mammal, and flowering plant fossils found in continental sediments.

Simulating Catastrophic Plate Tectonics

In 2003, Dr. John R. Baumgardner published an exciting paper[48] titled *Catastrophic Plate Tectonics: The Physics Behind the Genesis Flood*. This paper was exciting because, for the first time in the history of science, a breakthrough in computer simulation of catastrophic plate tectonics was achieved, as described in the paper. The simulation results validate the mechanism behind the Flood cataclysm that is presented above with both 2D and 3D simulations. Quoting Dr. Baumgardner—

"The crucial final piece of the puzzle has come from laboratory experiments that have carefully measured the way in which silicate

minerals deform under conditions of high temperature and high stress. These experiments reveal silicate material can weaken dramatically, by factors of a billion or more, at mantle temperatures and for stress conditions that can exist in the mantles of planets the size of the Earth. The scenario in which all the Earth's ocean lithosphere is rapidly recycled into the mantle via a runaway process, enabled by this stress-weakening behavior, is now known as catastrophic plate tectonics."

Also in the paper, Dr. Baumgardner describes how fundamental advances in computational techniques make it possible to advance the numerical solution successfully through the most extreme phase of the runaway phenomenon.

The 2D computer simulations covered in the paper (shown on the next two pages) simulate a thermal runaway instability within the earth's mantle, which is sized at approximately 1,800 miles (2,897 km) high (the actual thickness of the earth's mantle) and 7,200 miles (11,587 km) wide. Arrows indicate magma flow velocity scaled to the peak velocity (UMAX) in each Figure. Contour lines represent temperature in the top panel—and the base 10 logarithm of viscosity in the bottom panel—of each Figure. Initial temperatures are 300° Kelvin (K) at the top boundary and 2700° K at the bottom boundary. Initial viscosity at the middle of the mantle has a log value of 24 (i.e., 10^{24}). The first snapshot (Figure 15) simulates the mantle 5 days after the thermal runaway instability begins. The second snapshot (Figure 16) simulates the mantle after 12 ½ days have elapsed, and the third snapshot (Figure 17) simulates the mantle after 20 days. (Note: the top panel in each Figure shows visually the formation of magma (molten rock) *plumes,* which are the results of the runaway process. The reader may ignore the bottom panel in each Figure while focusing on the development of the plumes.)

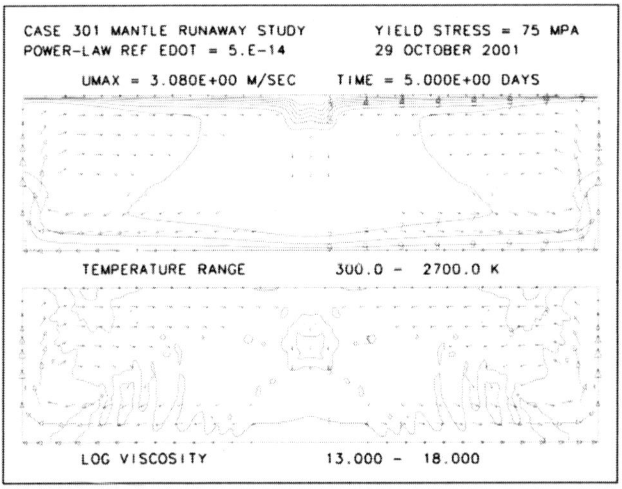

Figure 15—snapshot of thermal runaway instability within the earth's mantle 5 days after it began
courtesy Dr. John R. Baumgardner

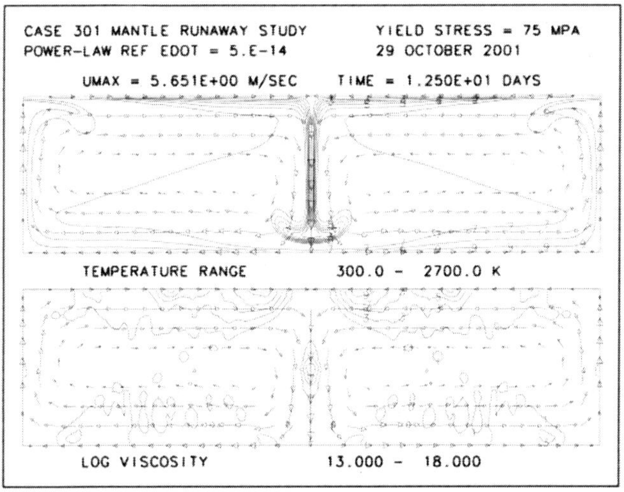

Figure 16—snapshot of thermal runaway instability within the earth's mantle 12.5 days after it began
courtesy Dr. John R. Baumgardner

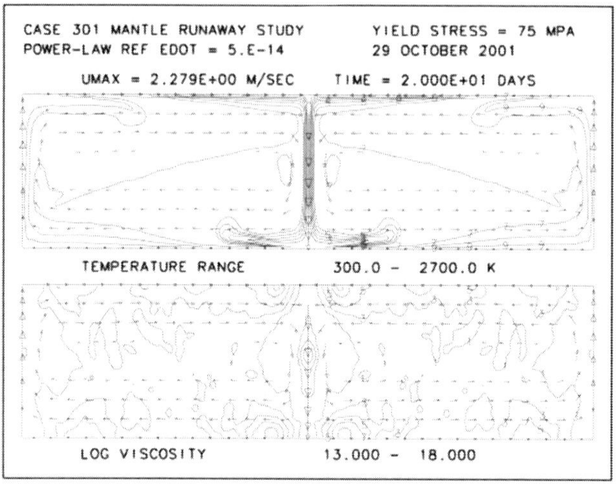

Figure 17—snapshot of thermal runaway instability within the earth's mantle 20 days after it began
courtesy Dr. John R. Baumgardner

Notice in Figure 15 (top panel) that the bottom boundary layer goes unstable first to produce a runaway *upwelling plume* of magma along each side of the mantle. These upwelling plumes in turn cause the top boundary layer to go unstable and also to runaway (Figure 16 top panel). Note in Figure 17 (top panel) that runaway plumes emerge from *both* top and bottom boundaries. The release of gravitational potential energy stored in both of these boundary layers drives the circular motion of the molten rock.

These hot magma upwellings (plumes) from the bottom of the mantle had dramatic implications on sea level during the Flood since they produced a temporary rise in the height of the ocean floor—by a mile (1.6 km) or more. Similarly, ocean plate *downwellings* from the top of the mantle caused a temporary depression of the mantle's upper boundary. Since the downwellings were usually located beneath continental regions, *they caused temporary depression of the continental surfaces as well (also by a mile/1.6 km or so).*

Since the viscosity (liquid "thickness") of the rock at the middle of the mantle initially had a magnitude of 10^{24} units, and since Figures 15-17 show runaway viscosities ranging from 10^{13} to 10^{18} units throughout the mantle (i.e., "LOG VISCOSITY 13.000 – 18.000" printed beneath each bottom panel), the mid-mantle viscosity (~10^{15} which is roughly half way between 10^{13} and 10^{18}) decreased by a factor of *one billion* (or 10^9—i.e., from 10^{24} to 10^{15}) after the thermal runaway instability began. However, deformational heating of the mantle shows no evidence of extreme temperatures associated with the runaway phenomenon. This is because the rate of deformational heating is proportional to viscosity, which diminished a billion-fold by the weakening associated with the runaway. Therefore, no damage to the earth's land masses occurred during the Flood as a result of extreme temperatures within its mantle.

The hot magma plumes simulated in Figures 15-17 would rise to the earth's surface and erupt there, producing large amounts of volcanic rock through fissure eruptions, such as, perhaps, the plateau basalts of South Africa, the Deccan Traps of India, and the Siberian basalts of Alaska and Canada. The correlation between plume formation and volcanic basalts has already been cited by geophysical scientists (for example in a paper by Stuart Weinstein of the University of Michigan titled "Catastrophic Overturn of the Earth's Mantle Driven by Multiple Phase Changes and Internal Heat Generation").

The author referred to catastrophic plate tectonics in the section above titled **Water within the earth**, where the following comment was recorded from an online article[74] posted by the University of Alberta—*"Knowing water exists beneath the crust has implications for the study of volcanism and plate tectonics, affecting how rock melts, cools and shifts below the crust."* The water within the earth's mantle undoubtedly contributed to the thermal runaway instability within the earth's mantle that Dr. Baumgardner was able to simulate.

Dr. Baumgardner also did a 3D spherical shell computer simulation of the Pangaea supercontinent fracturing that builds on the results of the 2D simulation summarized above. The initial shapes and extents of the tectonic plates in this simulation are shown in Figure 18 (see the next page). This starting point for the simulation represents an approximate reconstruction of Pangaea that is derived from the shapes of present-day continents and uses data taken from the present-day ocean floor. In order to initiate motion, a temperature drop of 400° K was introduced around most of the perimeter of the supercontinent—where the pre-Flood ocean lithosphere subduction first began, thus starting the process of breaking up the fountains of the great deep. (Note: when the ocean lithosphere first started subducting, it was much cooler—being at the bottom of the ocean—than the upper portion of the earth's mantle; hence lowering the temperature by 400° K along the perimeter of the pre-Flood supercontinent reflects what actually started the runaway phenomenon within the earth's mantle.)

Figure 18 shows how the continental blocks are being pulled apart, just 15 days after the simulation begins, due to the mantle runaway conditions that were modeled in the 2D calculations. While the resulting pattern of seafloor spreading and continent motion shown in Figure 18 is not identical to what can be inferred from today's earth, it is remarkably similar—especially given the lack of detail provided in the initial conditions.

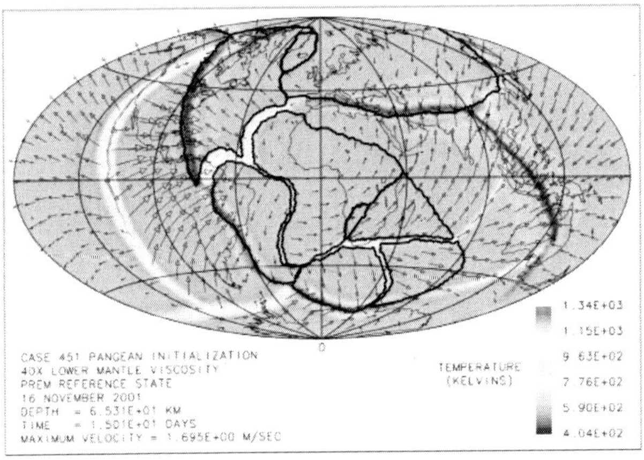

Figure 18—snapshot of the fracturing Pangaea supercontinent midway through the Flood (taken 15 days after the 3D simulation started)

courtesy Dr. John R. Baumgardner

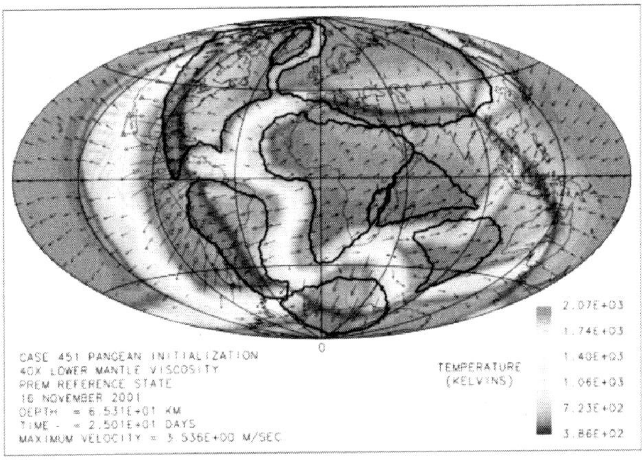

Figure 19—snapshot of 3D simulation results 10 days later, illustrating how the continents were starting to pull apart during the Flood

courtesy Dr. John R. Baumgardner

Since precise, observable data about the pre-Flood earth are simply not available, the 3D computer simulation shown in Figures 18 and 19 does not capture the earliest stages of the Flood cataclysm. However, the simulation does illustrate the type of catastrophic plate tectonics and mantle dynamics that actually occurred during the Flood. This simulation represents the earth's state roughly midway through the Flood, corresponding to the early Mesozoic timeframe in the geologic record (see Figure 9).

A comprehensive Flood simulation would ideally begin from an initial state resembling the pre-Flood earth, and the calculations would include the plate and mantle dynamics that unfolded during the Paleozoic portion of the geologic record (see Figure 9) as well as what followed afterward. This is, unfortunately, not possible—since no pre-Flood (i.e., Paleozoic) ocean floor still exists today! Even mainstream science confirms this.[47] Nevertheless, Dr. Baumgardner wrote this about his 3D simulation—

"I believe this calculation, even though it does not reach back to the very beginning of the actual cataclysm, nevertheless provides useful insight into the dynamics involved and reveals many details that otherwise might not be apparent. Hopefully, with sufficient effort it will be possible in the future to realize a pre-Paleozoic initial state suitably reliable to model the entire catastrophe."

The reader will undoubtedly be interested to learn that the planet Venus has been globally resurfaced in the relatively recent past by a catastrophic mechanism that was internal to the planet's mantle. High-resolution radar images from the NASA Magellan mission in the early 1990s led to this amazing discovery. More than half of the Venus surface was flooded with basaltic lava to produce mostly featureless plains, except for linear fractures caused by cooling and contraction. Recent runaway sinking of much of the planet's relatively cold upper thermal boundary layer into its mantle seems the most plausible mechanism to explain such a planetary resurfacing event. According to the authors of a 1994 paper [49] titled "The Global Resurfacing of Venus"—

"The impact cratering record on Venus is unique among the terrestrial planets. Fully 84% of the craters are in pristine condition, and only 12% are fractured. ... The constraints imposed by the cratering record strongly indicate that Venus experienced a global resurfacing event ... followed by a dramatic reduction of volcanism and tectonism. This global resurfacing event ended abruptly ... The present crater population has accumulated since then and remains largely intact."

Thus, the planet Earth is not the only planet in the Solar System to experience global, catastrophic plate tectonics!

At the end of his 2003 paper, Dr. John Baumgardner wrote—

"This is a historic moment. We have the key that unlocks secrets to the history of the Earth that no one has ever had before."

Utilizing this key—catastrophic plate tectonics—a clearer understanding of the Flood and its devastating effects emerges, as presented next.

Effect of the Flood on Living Things

The book of Genesis describes how the Flood began (highlights are the author's)—

In the six hundredth year of Noah's life, in the second month, the seventeenth day of the month, ***the same day were all the fountains of the great deep broken up, and the windows of heaven were opened. And the rain was upon the earth forty days and forty nights.*** (Genesis 7:11-12)

The impact of all the fountains of the great deep breaking open was a cataclysmic geologic event in and of itself, as discussed at length above, that restructured the entire planet. Gigantic waves of seawater, much larger and immeasurably more destructive than today's tsunamis, spread out in all directions from each spewing fountain, creating an ultra-destructive succession of colossal waves that drowned all land-based creatures and permanently transformed the land. Complex interactions between tectonic (earth's crust) and hydrodynamic (water in motion) forces caused the Flood waters to come and go in surges. The intensifying force of consecutive, pulsating waves pushed the water farther and farther inland, until eventually the entire supercontinent was immersed. The monstrous waves carried loose sediments and sea creatures with them, and deposited layer upon layer of this suffocating mixture on top of buried vegetation and land surfaces. The immense power of all this moving water was vastly greater than anything that has since been unleashed on earth.

At the same time, molten lava issued forth from each spouting fountain, causing the surrounding seawater to boil and then spray water vapor high into the atmosphere. This source of moisture fueled and refueled the windows of heaven so they could amplify and compound the torrential rain that was pouring from supersaturated clouds. Hypercanes, which were many times more destructive than present day hurricanes, made the tempests even more turbulent, adding to the chaos of the churning atmosphere. The fountains of the great deep essentially functioned as super-volcanos—with intensities that were multiple orders of magnitude beyond anything earth has experienced since.

While all this destruction was being inflicted by the violent, turbulent waters, subsurface shock waves of unimaginable magnitude were fracturing and setting in motion the tectonic plates that form the earth's crust. This mega-faulting resulted in massive earthquakes and volcanic eruptions that shook the whole earth and even impacted the earth's core. Magnetic stripes currently found in rocks that flank the mid-Atlantic Ridge rift valley—which was formed by tectonic plates moving away from each other—show that, during the Flood, the earth's magnetic field reversed its direction multiple times within a matter of days, thereby releasing energy from the earth's core at an accelerated rate.[50] As discussed earlier, this was caused by convective (swirling) motions within the earth's core, which convection was amplified by the vast amount of water contained within the core.

Try to imagine all the devastation described above going on for five long months without a pause! It took only forty days for the Flood waters to rise enough to bear up the ark, and then the destruction went on unabated for another 110 days! Genesis describes the horrors of this incomprehensible destruction as follows—

And the waters prevailed exceedingly upon the earth; and all the high hills, that were under the whole heaven, were covered. Fifteen cubits upward did the waters prevail; and the mountains were covered. And all flesh died that moved upon the earth, both of fowl, and of cattle, and of beast, and of every creeping thing that creepeth upon the earth, and every man. All in whose nostrils was the breath of life, of all that was in the dry land, died. And every living substance was destroyed which was upon the face of the ground, both man, and cattle, and the creeping things, and the fowl of the heaven; and they were destroyed from the earth: and Noah only remained alive, and they that were with him in the ark. And the waters prevailed upon the earth an hundred and fifty days. (Genesis 7:19-24)

The first 150 days of the Flood caused the eventual fossilization of trillions of *marine* (sea dwelling) organisms and the destruction of all *terrestrial* (land-based) and flying creatures. First to be impacted were those organisms dwelling on the bottom of the ocean, especially near the erupting fountains. Then organisms within the shallow continental shelf areas were overcome as the initial deadly waves and underwater surges inflicted their tremendous forces. When the huge waves broke over the coastline, anything dwelling there was smothered and eventually buried in the loose sediments that were transported by the raging sea. As the pulsing walls of water climbed ever higher and farther over the supercontinent, all life within the low-lying areas, and finally the upland regions, died a horrific death.

The Flood waters continued to rise and fall, and then rise again to new heights. Eventually the pre-Flood mountains (which are not the same mountains that exist today) were all covered. As the book of Genesis reports, the tallest mountain was covered by 22½ feet (15 cubits) of water. The height of that mountain is unknown, but is believed to have been lower than the stature of the taller mountains found on earth today, as discussed previously. Nevertheless, the pressure of the water that covered the supercontinent was extreme. For example, just the 22½ feet (6.9 m) of water above the tallest pre-Flood mountain would have increased the pressure on that mountain top by 10 pounds per square inch (29 kg/cm^2). This translates to almost 6½ tons per square yard (7050 kg/m^2). If a lower lying area had 1,000 feet (305 m) of water covering it, the increased pressure on that land surface would be nearly 285 tons per square yard (309,220 kg/m^2)! So the weight of the flood waters on the land added to the destruction that resulted from the moving waters.

After the 150 days of flooding had passed (highlights are the author's)—

... God made a wind to pass over the earth, and the waters assuaged. ***The fountains also of the deep and the windows of heaven were stopped, and the rain from heaven was restrained;*** *And the waters returned from off the earth continually: and after the end of the hundred and fifty days the waters were abated.* (Genesis 8:1-3)

But this was not yet the end of the Flood—

And the ark rested in the seventh month, on the seventeenth day of the month, upon the mountains of Ararat. And the waters decreased continually until the tenth month: in the tenth month, on the first day of the month, were the tops of the mountains seen. (Genesis 8:4-5)

And it came to pass in the six hundredth and first year, in the first month, the first day of the month, the waters were dried up from off the earth: and Noah removed the covering of the ark, and looked, and, behold, the face of the ground was dry. And in the second month, on the seven and twentieth day of the month, was the earth dried. (Genesis 8:13-14)

On the twenty seventh day of the second month, God told Noah and his family to leave the ark and release all the creatures that had been preserved during the Flood. This was 375 days after they had entered the ark. With immense relief and joy, Noah built an altar after departing the ark and offered burnt offerings to express his gratitude for the miraculous safety with which he and his cargo had been blessed for a year and ten days while riding out earth's greatest calamity.

Mount Ararat
(courtesy Henri Nissen, Wikimedia Commons)

Interestingly, the final resting place of the ark has been disputed by various cultures over time. Current Armenian and European traditions[51] identify the location as Mount Ararat in Turkey (near the Iran border in the easternmost part of Turkey). However, up until the 11th century, the Armenians associated the ark's landing place with Mount Judi, also in Turkey (south by southwest of Mount Ararat, near the headwaters of the Tigris River and the borders of Iraq and Syria). Jewish Babylonian, Islamic, and Syriac traditions also identify Mount Judi as the location. In the Qur'an 11:44, the final resting place of the vessel is called "Judi."

The author exposes this dispute because of the consequence it has on the depth of the Flood waters. Mount Judi is only 6,854 feet (2,089 m) high while Mount Ararat is currently 16,854 feet (5,137 m) above sea level. The Genesis record does not state that the ark rested on a mountain; rather, it rested upon "the mountains of Ararat." (Genesis 8:4)

Mount Ararat is the dormant cone of a stratovolcano, or composite volcano, that was built up with many layers (strata) of hardened lava. It has no volcanic crater, and has not been active in recent times. However, human bodies and artifacts from the Bronze Age (the 3rd Millennium BC) have been found under its hardened lava flows.[53] The Flood occurred during the 3rd Millennium BC, so one might speculate that Mount Ararat was somehow connected with the Flood, or with the post-Flood readjustment period of the earth's history. Interestingly, according to geologists[53] (highlights are the author's)—

"The formation of Ararat is hard to retrieve geologically, but the type of volcanism and the position of the volcano raise the idea that ***subduction relation volcanism occurred*** *..."*

If Mount Ararat existed during the Flood at an elevation of about 16,854 feet (5,137 m), then the amount of global Flood water required to cover its peak would be much, much greater than it would be if Mount Ararat did not even exist then, or was significantly lower than 16,854 feet (5,137 m) then, or if Mount Judi was the ark's final resting place. As discussed above, the topography of the pre-Flood earth was considerably flatter than it is now. Also, according to the catastrophic plate tectonics key presented above, volcanic activity was a direct consequence of the Flood. This key also reveals that the ocean floor was a mile (1.6 km) or so higher during the Flood than it is now, and that the supercontinent elevation may have been lowered by approximately a mile (1.6 km) (thus a two mile (3.2 km) reduction between them). In any case, the author is in awe at the amount of water and the magnitude of forces that were released by the Lord God Almighty to cleanse the earth by immersion in water.

There have been at least thirty reports claiming to have discovered Noah's ark. The author believes that one of these claims is, in fact, the remains of the ark—as shown in the picture below. The location of this discovery is in Turkey near the top of Doomsday Mountain, which is about 15 miles (24 km) south of Mount Ararat and just below Mount Judi. The elevation of this find, which is a National Treasure protected by the Turkish Government, is 6,375 feet (1,943 m) above sea level. In 1987, the site was designated by the Turkish Government as Noah's Ark National Park.

Noah's Ark National Park
A–single hull rib protrudes from the encasing mud
B–four hull ribs
C–collapsed cabin levels
D–ark's bow
Length from bow to stern is 300 Egyptian cubits (515 feet/157 m)
(courtesy viewzone.com)

The reasons why the author believes these are the remains of Noah's ark are beyond the scope of this book. However, the author encourages the reader to visit the following web sites to determine for himself/herself if Noah's ark has truly been found—

www.arkdiscovery.com/noah's_ark.htm
www.youtube.com/watch?v=i7iycpe16V0
www.viewzone.com/noahx.html

The sediments that were laid down one layer at a time by the ever-surging Flood waters contained organic debris. Over time, this debris either fossilized or *metamorphosed* (transformed) into coal, oil, and/or natural gas. A doctoral dissertation[54] by Steven A.

Austin in 1979 assembles compelling evidence that the world's major coal deposits are derived from huge mats of floating plant debris. Upon reflection, the author sees the foreknowledge of God at work here, as all this underground energy has fueled the amazing progress of modern society. Without it, life would be much more restrictive and difficult for God's children living on the earth today.

As mentioned previously, the earth's magnetic field was very dynamic during and after the Flood. The residual convective flow of the earth's molten outer core—amplified by the water contained within the core—resulted in more than two poles and frequent reversals of those poles. With such dynamic changes altering the earth's protective magnetic field, it is likely that the effectiveness of the field in shielding earth from cosmic radiation was severely compromised during this turbulent time. This undoubtedly resulted in severe genetic damage to the human genome, which likely contributed to one or both of the catastrophic reductions in lifespan that are highlighted in Figure 11. As the turbulence subsided, earth's magnetic field strength reached its post-Flood maximum about the time of Christ, and has been decaying ever since.

The Flood not only destroyed all living things on the face of the land, it also destroyed the earth's equilibrium—from the inner core to the land surfaces to the atmosphere high above. Naturally, the titanic forces that had been unleashed to carry out this devastation took time to subside. Furthermore, the imbalances they brought about continued to wreak havoc as readjustments of the various natural systems gradually took place. Hundreds of years were required for equilibrium to be reestablished. Residual catastrophes took place during this time, including the Division of the earth about a century after the Flood ended.

Also, the Flood's destabilization of climate patterns, ocean temperatures, and ocean currents led to the Ice Age, which lasted until the days of Abraham and Job (approximately 300 years after the Flood). The pre-Flood earth was warmer than today's earth, and the Flood warmed the oceans even more as the fountains of the great deep opened up and recycled the ocean's lithosphere. In addition, the Flood saturated the atmosphere with moisture. Land masses that greeted the sun after the Flood waters receded resembled mud slicks—which reflected much of the sunlight rather than absorbing it, since they had been covered by layers of new sediment and the top layers had just been eroded by the retreating waters. All these factors contributed to warm, wet winters after the Flood—ideal conditions for initiating the Ice Age. More snowfall (caused by more evaporation from the warmer oceans) and less snowmelt (caused by warmer average temperatures) allowed the ice to build up.

The post-Flood Ice Age formed earth's initial glaciers

Calculations show that the oceans' excessive heat took several hundred years to dissipate, and during this time the Ice Age dominated in the northern regions of the globe, as shown here.[55]

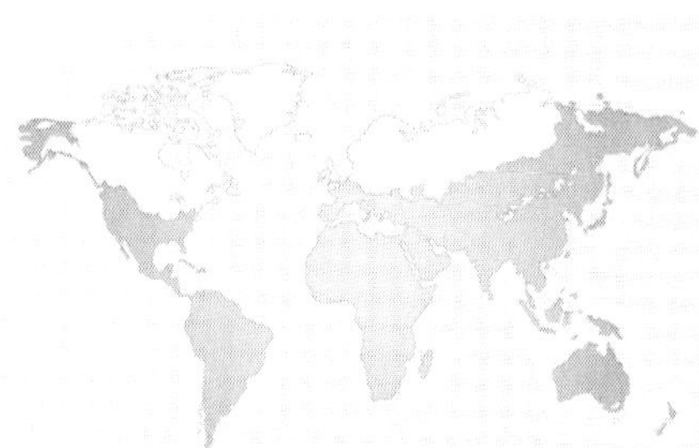

Extent of the Post-Flood Ice Age
(courtesy Anthony Watts)

Note these verses in the book of Job—

Out of whose womb came the ice? And the hoary frost of heaven, who hath gendered it? The waters are hid as with a stone, and the face of the deep is frozen. (Job 38:29-30)

Timing and Extent of the Division of the Earth

According to the computer simulation of catastrophic plate tectonics presented above, the pre-Flood supercontinent was divided into continents during the Flood—in other words, within 375 days and before the animals on the ark were released to *"breed abundantly in the earth, and be fruitful, and multiply upon the earth."* (Genesis 8:17)

The author does not believe that the continents actually moved into their current locations this quickly for a very simple but compelling reason—had the continents already separated by the time the land-based animals on the ark were released, there would have been no such native animal life on every continent—only on the continent on which the ark came to rest. How would kangaroos and platypuses have gotten to Australia? How would Kiwis have gotten to New Zealand? How would big cats have gotten to North and South America so they could evolve into cougars and jaguars there?

Obviously, the continents could not have moved into their present positions until a period of time had passed that gave land-based animal life enough time to reach every corner of the pre-Flood supercontinent. This period of time must have been relatively short, however, because every type of land-based animal is not native to every continent. For example, there are no snakes in New Zealand (even today!).

Therefore, the timing between the end of the Flood and the Division of the earth must have satisfied two constraints—enough time for *some* animals to reach every corner of the pre-Flood supercontinent, yet not so much time that *every* animal had time to get there.

The author believes that the century which passed between the end of the Flood and the birth of Peleg satisfies both constraints. Therefore, the author proposes that the literal Division of the earth truly did not happen until a century after the Flood ended, as recorded in Genesis. This implies that (*i*) the recycling of the ocean's lithosphere may have taken longer than the Creation Research Society has proposed, or that (*ii*) separation of the continents began with the Flood but continued for a century, or that (*iii*) catastrophic plate tectonics occurred twice—a century apart—with one cataclysm causing the Flood and the other dividing the continents and forming the mountain ranges that beautify the earth today. This latter concept—that catastrophic plate tectonics occurred twice—provides a plausible explanation as to why there are two catastrophic reductions in mankind's lifespan revealed in Figure 11.

Effects of the Flood on the Geologic Record

Perhaps the best way to try to grasp the effects of the Flood on the geologic record is to review the similar (but exceedingly miniaturized in comparison) effects of a modern volcano on its local geologic record.

On May 18, 1980, a magnitude 5.1 earthquake caused the summit and north face of Mount St. Helens in Washington State to start sliding downhill. One of the largest rockslides in recorded history, this avalanche traveled up to 155 miles per hour (249 km per hour) and crossed the western arm of nearby Spirit Lake. Most of the avalanche then traveled 13 miles (21 km) down the North Fork Toutle River, filling this river valley with up to 600 feet (183 m) of volcanic debris. About 24 square miles (62 km^2) were eventually covered.

Only seconds after it started to move, the rockslide also abruptly exposed the magma and high-pressure steam inside the volcano's neck to the much lower atmospheric pressure, causing a tremendous explosion. This explosion released energy equivalent to 20 million tons of TNT, which flattened 150 square miles (389 km^2) of forest in just six minutes. The deadly blast sent an extremely fast-moving current of hot gas, ash, and lava (collectively called *pyroclastic flow material*) shooting over the top of the slower moving rockslide. This *pyroclastic flow* moved initially at 220 miles per hour (350 km per hour), but it soon accelerated to 670 miles per hour (1,080 km per hour) and may have actually surpassed the speed of sound! As this deadly flow spread out, it devastated a fan-shaped area 23 miles wide and 19 miles long (37 by 31 km). In total, roughly 230 square miles (596 km^2) of forest were devastated, and the flow's extreme heat killed other trees miles away. Many of the trees that were caught in the flow were snapped off at their trunk bases.

Mount St. Helens eruption column
(courtesy US Geological Survey, Wikipeida Commons)

An eruption column, shown in the picture here, rose 15 miles (24 km) into the atmosphere and deposited ash in eleven different States. At the same time, sudden heating of the snow, ice, and glaciers that had capped the volcano's peak sent the melted ice racing down the northern slope. As the rushing water mixed with the pyroclastic flow materials mentioned above, the resultant flow became an extremely destructive *lahar*—or slurry of water, pyroclastic material, and rock fragments—that, with tsunami-like intensity, ravaged the hillsides and forest below. As the lahar reached rock strata in exposed areas, it quickly eroded that rock, and then deposited layered sediments in other areas. At the volcano's base, the sediment layers were up to 600 feet (183 m) thick.

Spirit Lake took the brunt of the volcano's fury. The aforementioned rockslide and the explosion it triggered splashed most of the lake's water, in the form of a colossal wave, against the northern mountain slopes to an astonishing height of 850 feet (260 m) above lake level. As the water ran back into its basin, it pulled with it millions of trees felled by the pyroclastic flow mentioned above. These trees formed a floating log mat that covered about 40% of the lake's surface (2 square miles or 5 km^2). A picture of this log mat, shown here, was taken 32 years after Mount St. Helens first erupted. The rockslide also brought with it 350,000 acre-feet of *pyrolized* trees (turned to char by intense heat) and other volcanic debris, suddenly reducing the lake's volume. The lahar and pyroclastic flow deposits blocked the lake's outlet into the North Fork Toutle River valley, raising the surface of the lake by approximately 200 feet (60 m). This nearly doubled the surface area of the lake.[56, 57]

Spirit Lake log mat in 2012 *(courtesy Stephan Schulz, Wikipeida Commons)*

The total energy output of this catastrophic volcano eruption on May 18 alone was equivalent to 400 million tons of TNT—which is also equivalent to 20,000 atomic bombs having the same destructive potential as those dropped on Hiroshima and Nagasaki at the end of World War II. Consequently, Mount St. Helens became *"a miniature laboratory for catastrophism."* [58]

Turning now to the geologic record written by Mount St. Helens starting May 18, 1980, the layers of sediment up to 600 feet (183 m) thick that were deposited at the volcano's base hardened into fairly solid rock in just a few years. Today, these layered sediments look very similar to other rock strata dated at millions of years.

According to Dr. Steven Austin, who personally investigated[58] the geologic changes that occurred as a result of the volcano,—

"Mudflows ... were responsible for the most significant erosion. A mudflow on March 19, 1982, eroded a canyon system up to 140 feet [43 m] deep in the headwaters of the North Fork of the Toutle River..." This *"little 'Grand Canyon of the Toutle River' is a one-fortieth scale model of the real Grand Canyon. The small creeks which flow through the headwaters of the Toutle River today might seem, by present appearances, to have carved these canyons very slowly over a long time period, except for the fact that the erosion was observed to have occurred rapidly!"*

How amazing that such a canyon could be carved out of stone in a single afternoon!

The log mat floating on Spirit Lake (shown on the previous page) consists of approximately four million trees that had their bark and branches stripped away by wind and waves. Prior to August 1986, scuba divers had discovered water-saturated sheets of tree bark on the bottom of the lake below this log mat. This bark has already morphed into a layer of *peat* (partially decayed plant material) several inches thick. The peat layer resembles, both compositionally and texturally, many of the coal beds in the eastern United States—which also formed from tree bark and appear to have accumulated beneath floating log mats. Thus, the Spirit Lake peat is thought to be in the first stage of coal formation. It only needs to be buried by another eruption and heated slightly to transform into a modern coal bed.

Once again, how amazing that this layer of peat was created in such a short time—less than six years after Mount St. Helens erupted!

In addition to the horizontal trees that comprise the log mat floating on Spirit Lake, more than 15,000 logs were observed in 1985 to be floating in an upright position, with root ball submerged. The average height of these vertical trees is twenty feet (six m). Some have the upper portion of their trunk above the water while others are entirely submerged. For some of those submerged, their trunk bases are already buried in more than three feet (one m) of sedimentation. Other trees are not buried as such, indicating that Spirit Lake's upright trees were deposited at different times. If all these vertical trees were found buried in the geologic record, they would likely be interpreted as having come from multiple forests which grew on different levels over thousands and thousands of years—just as the petrified upright forests of Yellowstone National Park have been explained.[58]

And yet, the vertical trees in Spirit Lake were all deposited in just a few short years!

Soon after the eruption of Mount St. Helens in 1980, the waters of Spirit Lake became contaminated from volcanic gases seeping up from the lake bed. By the time a month had passed, the water was virtually devoid of oxygen. Scientists predicted that it would take a very long time for Spirit Lake to recover, but the re-emergence of *phytoplankton* (photosynthesizing microscopic organisms) in 1983 started the process of restoring oxygen levels. Before long, frogs, salamanders, and fish (reintroduced by fishermen) were thriving. This is another example of natural processes occurring much, much faster than would be expected through the eyes of uniformitarianism.[57]

Having reviewed the astonishing, calamitous effects that a single erupting volcano spewed to rewrite its local geologic record in a time span of just minutes to a few short years, the reader should now attempt to visualize what the interminably more powerful global Flood accomplished during the full year of its destructive reign on earth.

No person, geologist or otherwise, can correctly interpret the geological record anywhere without considering the Flood and the catastrophic plate tectonics that powered it—for together, they inflicted every region of earth's lands and seas with horrendous forces that caused the then-existing world to perish.

Because of the Flood, the geologic record of our present world must now be seen for what it really is—the aftermath of an unparalleled, worldwide calamity that occurred about 4,500 years ago.

To assist with this visualization, it is helpful to understand when, in the geologic timescale, the Flood began. Figure 9 is reproduced on the next page for easier reference while considering this question.

As stated previously, fossils are found only in the rock layers above the Hadean layer. The fossil record indicates that, prior to 580 million years ago (during the Archean and Proterozoic eons), most organisms were extremely simple—just individual cells that were sometimes grouped into colonies. According to mainstream science, the rate of evolution accelerated greatly between 580 and 500 million years ago, and the diversity of life began to resemble that of today's plants and animals.

Notice in Figure 9 that a "Cambrian Explosion" took place about 530 million years ago. This geologic and evolutionary milestone was so named because it represents the relatively rapid appearance of most marine animal life found in the fossil record—without any trace of fossilized predecessors anywhere else in the fossil record. Even today, the Cambrian Explosion is considered a mystery and is the subject of much debate.[59]

When Charles Darwin published *On the Origin of Species by Means of Natural Selection* in 1859, he was concerned that this sudden appearance of numerous different animal groups in the fossil record, with no evidence of evolution from more primitive ancestors, could be raised as a serious objection to his theory of evolution. His personal belief was that the oceans had teemed with living creatures before this point in the geological timeframe, but their fossils were missing because of deficiencies in the fossil record.[60]

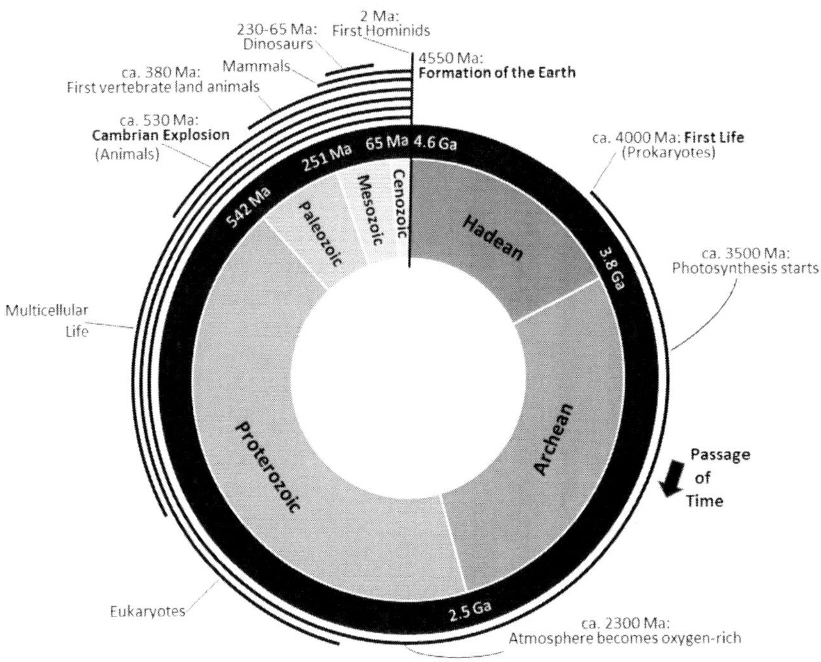

Figure 9—Geologic Time Scale Shown as a Clock
(Ma = 1 million years, Ga = 1 billion years)

Later on, the American paleontologist Charles Walcott, who became director of the U.S. Geological Survey in 1894, suggested that the "Lipalian" interval of the geological clock (not shown in Figure 9, but represents the time interval between the youngest [i.e., shallowest] Precambrian and the oldest [i.e., deepest] Cambrian sediments) was the period of time when ancestors of Cambrian animals evolved, but that no fossil record of them was preserved. He proposed that the Lipalian rocks have either been eroded or they remain undiscovered. Thus Walcott corroborated Darwin's opinion. However, mainstream geology never adopted Walcott's hypothesis.

The author considers this proposal of Walcott's to be one of Dr. Richard Lewontin's "absurd constructs" and an "unsubstantiated just-so story."

According to the online Encyclopedia Britannica[61] (highlights are the author's)—*"Walcott was supported in his views by the fact that* ***an almost worldwide Precambrian-Cambrian unconformity does indeed exist****. Even more remarkable is the clear distinction between the virtually barren Precambrian horizons and the productive Cambrian strata, because* ***many localities are now recognized throughout the world in which no break exists in the sedimentary record from the latest Precambrian*** *(sometimes termed Eocambrian)* ***to the earliest Cambrian.****"*

The *"almost worldwide Precambrian-Cambrian unconformity"* mentioned above is known as the "Great Unconformity." It was named by the explorer John Wesley Powell in 1869 during the first documented exploration of the Grand Canyon. According to modern geology, an *unconformity* is a gap in the geological record where time appears to be missing between rock strata. Such an unconformity is identified when a layer of sediment lies upon what is considered a much older geologic layer—thus a gap in time between them is assumed to be missing. The Precambrian-Cambrian unconformity consists of Cambrian sediments (assumed to be on the order of 525 million years old) sitting atop Precambrian rocks (assumed to be on the order of 1.5 billion years old). Figure 20, shown here, illustrates the Great Unconformity in the Grand Canyon that was discovered by Powell.[62]

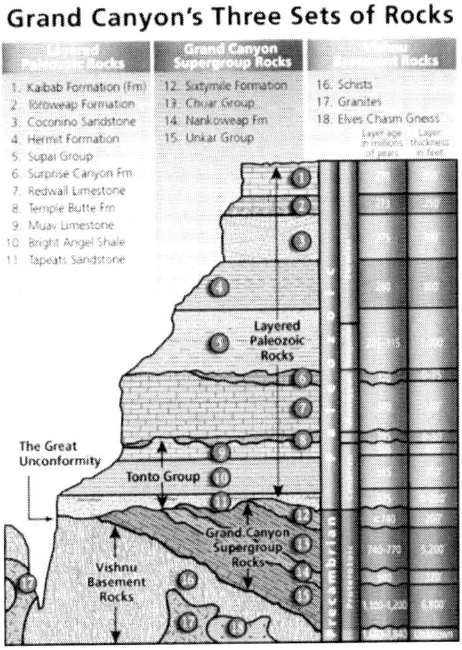

Figure 20—Grand Canyon's rock layers, showing the Great Unconformity

(*courtesy U.S. Department of the Interior, Wikimedia Commons*) (*in the public domain*)

Of course, modern science has developed several different hypotheses, in addition to Walcott's, to explain away the Cambrian Explosion mystery. One recent proposal is that the weathering of rock, as inferred from the Great Unconformity, dumped a chemically altered type of sediment into the ocean that actually triggered the Cambrian Explosion. The authors of this paper [63] wrote—

"Although Darwin and others have interpreted this widespread hiatus in sedimentation on the continents as a failure of the geologic record, this palaeogeomorphic surface represents a unique physical environmental boundary condition that affected seawater chemistry during a time of profound expansion of shallow marine habitats. Thus, the formation of the Great Unconformity may have been an environmental trigger for the evolution of biomineralization and the 'Cambrian explosion' of ecologic and taxonomic diversity following the ... emergence of animals."

A November 2010 news release from the University of California at Santa Barbara (UCSB) eliminated the Cambrian Explosion mystery altogether by simply reclassifying the early Cambrian rocks as Precambrian rocks! According to Susannah Porter, associate professor [64] in the Department of Earth Science at UCSB—

"We found that with improved dating and correlation of rock sequences, the short burst of appearances goes away. Instead, appearances of the earliest skeleton-forming animals were drawn out over more than 20 million years."

The author considers both of these proposals to belong to Dr. Lewontin's categories of "absurd constructs" and "unsubstantiated just-so stories."

Surely the reader can see a simpler, more logical, and more truthful explanation for the Cambrian Explosion and the Great Unconformity!

Chapter 2 of this book has already provided scientific proof that Darwinian evolution does not occur at all, and this chapter has presented breakthrough science that (*i*) exposes the unreliability of radioactive dating methods and (*ii*) explains how the Flood would have caused both the Cambrian Explosion and the Great Unconformity. Before the Flood, human beings lived for nearly a millennium, so it is logical to assume that antediluvian animals also lived much longer than they do now. Furthermore, the earth experienced no catastrophic geologic events after the Fall and before the Flood, hence one would not expect to see a geologic record rich in animal fossils prior to the Flood.

Tightly folded sedimentary strata near Rock Springs, Wyoming must have been soft when bent. Only the forces of a global flood could have formed such a remarkable geologic wonder.

To be clear, the Flood was responsible for the sedimentary strata that contain abundant animal fossils, since all animal life was destroyed by the Flood. These strata were deposited upon antediluvian rock by the Flood and they comprise the geologic record for the Paleozoic, Mesozoic, and Cenozoic eras (see Figure 9).

As Dr. John Baumgardner [65] explained—

"... the beginning of the Flood must correlate with the abrupt discontinuity in the fossil record at the Precambrian-Cambrian boundary. This point in the geological record not only marks a sharply-defined beginning to the abundant occurrence of fossils, but it also represents a nearly worldwide stratigraphic unconformity in the rocks themselves."

The reader will be interested to know that the igneous (solidified magma), metamorphic (formed by extreme heat and pressure), and sedimentary rocks immediately below the Great Unconformity in the Grand Canyon have been *"beveled off to an amazingly flat surface. Except for a few places where a hard, resistant Shinumo boulder is lodged near the erosion surface, the erosion plane exhibits little relief."* [66] Furthermore, the rock layer directly above the Great Unconformity consists of boulders the size of bushel baskets along with very large cobblestones.

How could the lower layer of the Great Unconformity exhibit extensive sheet erosion while the layer above it be composed of clustered cobbles and large boulders—and all of this cover a semi-continental area (encompassing most of the United States, western Canada, and over to northern Greenland)? The obvious answer is the Flood. Only a catastrophic, global Flood could wreak such an extensive result. As the author of a 2011 paper about this subject (the Tapeats Sandstone) wrote at the end of his paper,[67]

"Thus the Tapeats is easily interpreted within the constraints of the Genesis Flood but less easily so by uniformitarian or actualistic models."

Additional evidence that the Flood was responsible for the sedimentary strata in the geologic record involves the huge areal extents of many sedimentary deposits.

Most sedimentary rock found on the continents today consists of fine-grained shale, mudstone, and siltstone. A common black shale in North America is called the Chattanooga Shale. First discovered in Chattanooga, Tennessee, this widespread shale is found in northern Mexico; the Canadian Provinces of Ontario, Manitoba, Saskatchewan, Alberta, and British Columbia; and also the following States—

1. Maine
2. New Hampshire
3. Vermont
4. New York
5. Pennsylvania
6. Maryland
7. Virginia
8. West Virginia
9. Ohio
10. Indiana
11. Michigan
12. Wisconsin
13. Minnesota
14. Illinois
15. Kentucky
16. Tennessee
17. Mississippi
18. North Carolina
19. Georgia
20. Alabama
21. Missouri
22. Arkansas
23. Iowa
24. Texas
25. New Mexico
26. Oklahoma
27. Kansas
28. Nebraska
29. South Dakota
30. North Dakota
31. Montana
32. Wyoming
33. Idaho

Shale sample
courtesy U.S. Geological Survey, Mineral Information Institute, Wikimedia Commons (in the public domain)

Rarely more than 20 feet (6 m) thick, the Chattanooga Shale lies on top of an amazingly flat and featureless surface that was formed by rapid sheet erosion. The Shale consists of countless fine laminations (layers) which can be easily separated. Fossils of marine creatures and land plants are frequently found between the layers, and burrows formed by sea creatures trying to escape an underwater grave are readily apparent. The laminations provide evidence of *hummocky* (wave-rippled) cross-stratification, which results from rapidly flowing water and continuous particle deposition. All these characteristics provide compelling evidence of a global Flood. No modern geologic analogy exists for such an extensive deposit.

Another sedimentary rock stratum that has an enormous areal extent is the St. Peter Sandstone—a cross-bedded deposit of sandstone that covers more than 1.5 million square miles (3.88 million km^2) of North America with a thickness ranging between 100 and 300 feet (30 to 90 m). This sandstone is found in all the Countries, Provinces, and States identified above for the Chattanooga Shale—plus the States of Colorado, Utah, Arizona, and Nevada. This stratum extends about 2,000 miles (3,200 km) from east to west, and from Mexico to Canada. Depositing such a thick layer of sandstone required the swift movement of deep water and an inconceivable amount of sand—nothing short of the global Flood could have formed this enormous geologic wonder.

A third example of a sedimentary rock stratum that can only be attributed to the Flood is the dinosaur fossil-bearing Morrison Formation, which covers the States of Montana, Wyoming, and Colorado, and also extends into the States of North and South Dakota, Idaho, Nebraska, Kansas, Utah, Arizona, and New

Mexico. This Formation is about 100 feet (30 m) thick on average and covers an area of about 600,000 square miles (1.55 million km^2). The popular Dinosaur National Monument Quarry in northeastern Utah allows visitors to view an exposed wall of the Morrison Formation that reveals nearly 1,500 dinosaur bones. Contained within the Morrison Formation are—

"... thousands of partial dinosaur carcasses, and tens of thousands of shattered dinosaur bones. Obviously, the articulated skeletons required rapid burial, for scavengers and bacteria are normally ever-present to devour any unprotected remains. Interestingly, the most common fossil at Dinosaur National Monument in the Morrison Formation is that of an articulated clam."

Animals and plants from a variety of environments—ocean, land, mountainous, and coastal—are all mixed together as fossils. Therefore, this geologic record is not a snapshot of real life in the past; rather, it is a record written in stone of violent death and extensive transportation.[68]

A fourth example of evidence for the global Flood is the Shinarump Conglomerate—a deposit in the southwestern United States that covers about 125,000 square miles (324,000 km^2) at an average thickness of 50 feet (15 m). Unlike the deposits discussed above, the Shinarump Conglomerate contains plentiful petrified wood mixed with large boulders and hefty cobbles. In order to spread such massive rocks with the remains of an extensive forest over an area roughly 1,300 miles (2,100 km) long and 350 miles (560 km) wide, rapid water movement over a regional scale was required. No such uniformitarianism water force exists today.

While the four examples given here are located in North America, the reader should not assume that the only compelling evidence for the global Flood exists in North America. On the contrary, geologic evidence of the destructive work of the global Flood is found on every continent.

One example is a set of formations that comprise the Germanic Trias Supergroup (*Germanische Trias-Supergruppe*)—the Keuper, Muschelkalk, and Bunter. Similar formations, which have nearly identical coloration and physical properties compared to the Germanic Trias Supergroup, are also found across Europe, from England to Bulgaria, and in North America along the eastern seaboard and in Texas, New Mexico, and Arizona. Also, a distinctive geologic conglomerate of purple and white quartzite boulders located in England is found throughout Europe, with particularly fine examples in France, Spain, and Bulgaria.[69]

In conclusion, had the Global Flood not occurred, there would be no sedimentary strata to comprise earth's geologic record for the Paleozoic, Mesozoic, and Cenozoic eras. Nor would there be a fossil record within these strata, as will be shown in the next chapter.

CHAPTER FOUR

The Fossil Record

Contrary to popular belief, the fossil record does not support the theory of Darwinian evolution. In fact, breakthrough science is challenging old thinking about the fossil record!

The Fossil Record According to Mainstream Science

In Chapter 3, the author introduced the high level summary of the geologic time scale shown on page 185.

Chapter 3 also pointed out the following—

1. Rocks corresponding to the Hadean eon contain no fossils; however, fossils are found in the rock layers directly above the rocks in the Hadean layer.
2. The other eons—Archean, Proterozoic, and Phanerozoic (which is not shown)—encompass the evolution of life. Together, the Paleozoic, Mesozoic, and Cenozoic eras form the Phanerozoic eon.
3. A two million year Quaternary period, which is the time that recognizable humans have been on the earth, is too small to be visible at the scale of this clock.
4. The circular lines that partially circumscribe the geologic clock in Figure 9 indicate how long different types of living things existed on the earth, as revealed in the fossil record. For example, dinosaur fossils are found in the strata corresponding to the Mesozoic era.

5. No fossil samples have been found anywhere in the world that have a $^{14}C/C$ ratio as low as 0.001 pmc, and therefore they are all less than 90,000 years old.
6. Since most fossils seem to have approximately the same $^{14}C/C$ ratio, it is plausible that all the fossilized organisms lived on the earth at the same time.
7. Modern fossil dating does not agree with the geologic time scale.
8. The fossil record indicates that, prior to 580 million years ago (during the Archean and Proterozoic eons), most organisms were extremely simple—just individual cells that were sometimes grouped into colonies.
9. According to mainstream science, the rate of evolution accelerated greatly between 580 and 500 million years ago, and the diversity of life began to resemble that of today's plants and animals.
10. A "Cambrian Explosion" took place about 530 million years ago. This geologic and evolutionary milestone was so named because it represents the relatively rapid appearance of most marine animal life found in the fossil record—without any trace of fossilized predecessors anywhere else in the fossil record.
11. Darwin was concerned that the sudden appearance of numerous different animal groups in the fossil record [the Cambrian Explosion], with no evidence of evolution from more primitive ancestors, could be raised as a serious objection to his theory of evolution.
12. Walcott suggested that the Lipalian interval of the geological clock (not shown in Figure 9, but represents the time interval between the youngest Precambrian and oldest Cambrian sediments) was the period of time when ancestors of

Cambrian animals evolved, but that no fossil record of them was preserved.

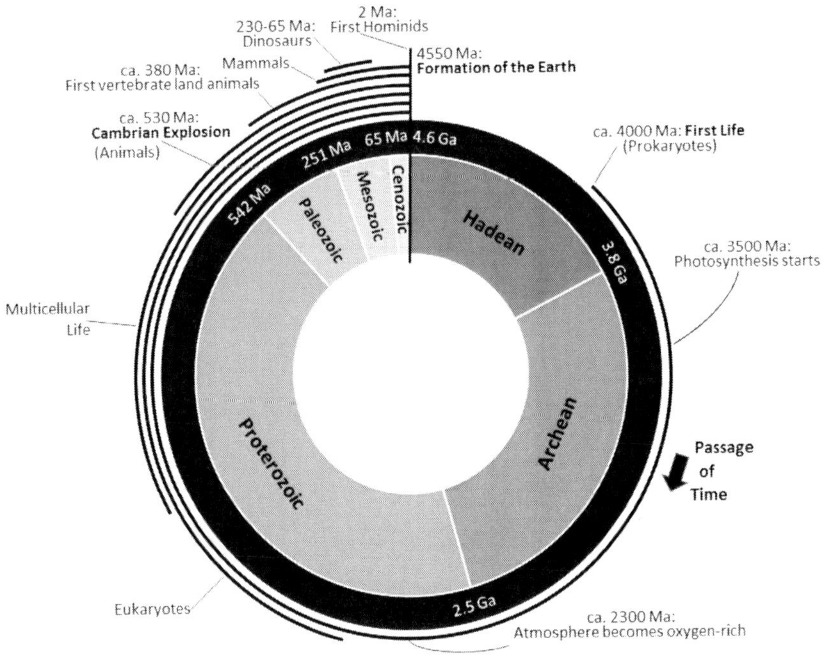

Figure 9—Geologic Time Scale Shown as a Clock
(Ma = 1 million years, Ga = 1 billion years)

The issue raised by Darwin and Walcott (covered in paragraphs 10 through 12 above) introduces a major problem that the fossil record creates for mainstream science. The problem is that **essentially no trace of fossilized predecessors can be found anywhere**—not only just before the Cambrian Explosion.

To get around this problem, the concept of *punctuated equilibria* was introduced by Dr. Niles Eldredge and Dr. Stephen Gould in 1972.[1] Their theory proposed that, when major evolutionary change occurs, a species rapidly splits into two distinct species rather than transforming gradually into a different single species. Eldredge and Gould admitted that gradual evolution of the species is virtually nonexistent in the fossil record. They also acknowledged that the fossil record shows that *stasis* (little or no evolutionary change) characterizes the history of most fossil species. In a 1977 essay[2] titled "Evolution's Erratic Pace," the late Dr. Gould, a Harvard paleontologist, wrote the following—

"The history of most fossil species includes two features particularly inconsistent with gradualism: 1. Stasis—most species exhibit no directional change during their tenure on earth. They appear in the fossil record looking much the same as when they disappear; morphological change is usually limited and directionless. 2. Sudden appearance—in any local area, a species does not arise gradually by the steady transformation of its ancestors; it appears all at once and 'fully formed.'"

Of course, Dr. Gould did not see his ideas contradicting Darwin; in fact, he claims they are the outcome of what Darwinian evolution predicts, as he expressed in the same essay—

"Species should be static through their range because our fossils are the products of large central populations. In any local area inhabited by ancestors, a descendant species should appear suddenly by migration from the peripheral region in which it evolved. In the peripheral region itself, we might find direct evidence of speciation, but such good fortune would be rare indeed because the event occurs so rapidly in such a small population. Thus, the fossil record is a faithful rendering of what evolutionary theory predicts, not a pitiful vestige of a once bountiful tale."

In another essay[3] written for the magazine *Natural History*, Gould added this (highlights are the author's, not Gould's)—

"All paleontologists know that the fossil record contains precious little in the way of intermediate forms; transitions between major groups are characteristically abrupt.

*"**Even though we have no direct evidence for smooth transitions, can we invent a reasonable sequence of intermediate forms, that is, viable functioning organisms, between ancestors and descendants?** Of what possible use are the imperfect incipient stages of useful structures? What good is half a jaw or half a wing?"*

Does this not smack of Dr. Lewontin's "unsubstantiated just-so stories?"

Eldredge and Gould were not the only scientists to acknowledge the lack of gradual evolution in the fossil record. Dr. Steven Stanley wrote the following in his book[4] *Macroevolution*—

"The known fossil record fails to document a single example of phyletic evolution accomplishing a major morphological transition [involving a basic body style change] *and hence offers no evidence that the gradualistic model can be valid."*

Others have made similar acknowledgements.

Not surprisingly, geneticists did not accept the concept of punctuated equilibria because they could not envision any way that genetic mutations could result in rapid, coordinated transitions of soft and hard body parts. The section Creation of a New Gene in Chapter 2 confirms why these geneticists were reluctant to accept this attempt to sidestep the major problem that the fossil record causes for evolutionists (other than the origin of life itself). Of course, paleontologists who had been struggling with the lack of evolutionary transitions in the fossil record were very quick to accept punctuated equilibria. Therefore, paleontologists and geneticists argued about punctuated equilibria for several years, but the arguing subsided once evolutionists stopped using the fossil record as evidence for evolution. Note the words[5] of evolutionist Mark Ridley in 1981—

"No real evolutionist, whether gradualist or punctuationist, uses the fossil record as evidence in favor of the theory of evolution as opposed to special creation ..."

The author finds it astonishing that evolutionists have turned away from the fossil record in order to justify their theories! On what, then, is modern evolution theory founded?

No wonder Dr. Richard Lewontin wrote[6] (emphases are the author's, not Lewontin's)—

"We take the side of science in spite of the ***patent absurdity of some of its constructs****, in spite of its failure to fulfill many of its extravagant promises of health and life, in spite of the* ***tolerance of the scientific community for unsubstantiated just-so stories****, because we have a prior commitment to materialism ...* ***We are forced by our a priori adherence to material causes to create an apparatus of investigation and a set of concepts that produce material explanations, no matter how counterintuitive, no matter how mystifying to the uninitiated.*** *Moreover, that materialism is absolute, for we cannot allow a Divine Foot in the door."*

As recently as 2007, Dr. Eugene Koonin, a senior investigator at the National Center for Biotechnical Information, confirmed[7] this fundamental problem that mainstream science faces—

"Major transitions in biological evolution show the same pattern of sudden emergence of diverse forms at a new level of complexity. The relationships between major groups within an emergent new class of biological entities are hard to decipher and do not seem to fit the tree pattern that, following Darwin's original proposal, remains the dominant description of biological evolution."

The reader may be interested to learn what Dr. Koonin's solution[7] to the problem is—

"I propose that most or all major evolutionary transitions that show the 'explosive' pattern of emergence of new types of biological entities

correspond to a boundary between two qualitatively distinct evolutionary phases. The first, inflationary phase is characterized by extremely rapid evolution driven by various processes of genetic information exchange, such as horizontal gene transfer, recombination, fusion, fission, and spread of mobile elements. These processes give rise to a vast diversity of forms from which the main classes of entities at the new level of complexity emerge independently, through a sampling process. In the second phase, evolution dramatically slows down, the respective process of genetic information exchange tapers off, and multiple lineages of the new type of entities emerge, each of them evolving in a tree-like fashion from that point on.

"This biphasic model of evolution is substantially analogous to the scenario of the origin of universes in the eternal inflation version of modern cosmology. Under this model, universes like ours emerge in the infinite multiverse when the eternal process of exponential expansion, known as inflation, ceases in a particular region as a result of false vacuum decay, a first order phase transition process. The result is the nucleation of a new universe, which is traditionally denoted Big Bang, although this scenario is radically different from the Big Bang of the traditional model of an expanding universe. Hence I denote the phase transitions at the end of each inflationary epoch in the history of life Biological Big Bangs (BBB)."

In order to justify the first, inflationary phase of Koonin's model, he added[7] this—

"It is proposed that bacterial and archaeal phyla emerged independently from two distinct populations of primordial cells that, originally, possessed leaky membranes, which made the cells prone to rampant gene exchange ..."

Where is the scientific evidence for this proposal? How could anyone on earth today know how "primordial cells" functioned—or that they even existed? The author presents this BBB proposal as an example of the type of "unsubstantiated just-so stories" upon which modern evolution theory is founded.

Remarkably, the fossil record is no longer mainstream science's foundation for scientific evidence!

More recently, the use of *cladistics* has replaced fossil searching as the tool[8] of choice for evolutionists.

"Cladistics is an approach to biological classification in which organisms are grouped together based on whether or not they have one or more shared unique characteristics that come from the group's last common ancestor and are not present in more distant ancestors. Therefore, members of the same group are thought to share a common history and are considered to be more closely related."

Willi Hennig, a German biologist, was the founder of the cladistics method, although its first use appeared in 1901, twelve years before he was born. Hennig referred to his work as "phylogenetic systematics" (the title of his 1966 book), but evolution researchers have since popularized the term cladistics. The result of a cladistic analysis is a *cladogram*—a tree-shaped diagram that is assumed to represent the best hypotheses of evolutionary relationships. An example of a horizontal cladogram is shown here. Vertical cladograms, with the ancestor at the bottom, are also popular.

Horizontal Cladogram Example

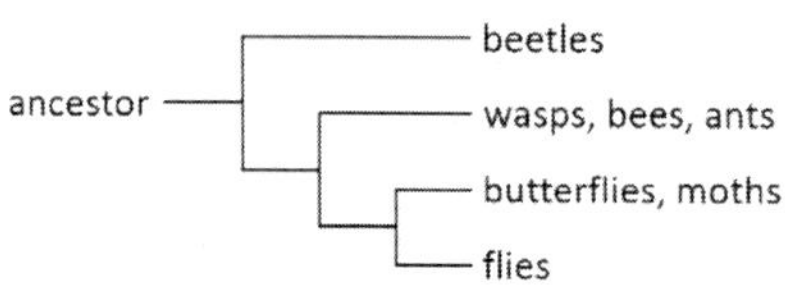

While such cladograms were originally calculated by hand and were based on the high level form and structure of organisms, modern techniques utilize molecular DNA sequencing and computerized algorithms to arrive at a result. Nevertheless, the results vary by author—since disagreement about which traits are considered the most pertinent are common.

For example, Michael Benton's latest textbook *Vertebrate Palaeontology* (third edition) presents seven competing cladograms regarding the evolution of fish to *tetrapods* (the first four-limbed vertebrates). These competing cladograms represent four basic schools of thought[9] by various evolutionists. Furthermore, cladistics has been criticized from its beginning for using circular reasoning and subjective judgments, as just exemplified.[10]

Cladograms should be recognized for what they really are—graphical illustrations of an individual researcher's idea—not fact, and certainly not proof of evolution!

In the author's opinion, the science of modern evolution has become, in reality, pseudoscience—for it is not built on a firm scientific foundation. This should not be surprising to the reader, however, since Chapter 2 proved that forward evolution is false.

Unlike mainstream evolution scientists, creation scientists consider the fossil record to be crucial for obtaining truthful scientific evidence of life's history on earth. The fossil record does not lie. However, different interpretations of the fossil record lead to different theories and hypotheses. These proposals from both mainstream and creation scientists cannot all be true, however, because they are typically incompatible with one another. Fortunately, scientific work regarding the fossil record has been performed by members of the Creation Research Society, who are trained experts, and who are not ideologically or religiously committed to Darwinian evolution or uniformitarianism. The remainder of this chapter will review their work, as summarized in *The Fossil Record—Unearthing Nature's History of Life* by Frank J. Sherman and John D. Morris.[11] The author of this book recommends *The Fossil Record* to readers who desire more detail than can be provided here.

Fossil Record Basics

Fossil origins

Normally, when a living organism dies—plant or animal—its physical remains begin to decay. The remains may also be consumed by scavengers. Anything the scavengers leave is either decomposed by bacteria or disintegrates from oxidation. Eventually, everything that composed the physical remains is gone. Thus, no fossil remains are created.

American bison (buffalo)
courtesy U.S. Dept. of Agriculture, Wikimedia Commons (in the public domain)

To illustrate, consider the American bison, or buffalo. In the early 1800s, millions of these large animals roamed the prairies of North America. When the United States Government ordered the slaughter of buffalo in the 1870s, millions of carcasses soon littered the American prairie. Yet no fossils of these animals are found today—except for a few skulls and teeth that were rapidly buried by small floods. All the other remains were consumed by decay, scavengers, bacteria, and/or oxidation—including bones.

In order for a fossil to be created, an organism must be encapsulated in tree sap, mud, or a liquid sediment very soon after it expires, or even *before* it dies. Then the organism, alive or dead, must remain intact in its liquid tomb until the tomb becomes part of the geologic record. Such a process is the only way that the natural onset of organic decomposition and disintegration can be prevented.

Fossil types

- Unaltered Hard Parts—most animal fossils are marine (oceanic) shellfish, whose hard shells remain intact over time with their original chemical composition. This fossil category also includes bone, coral, and wood fossils that have maintained their composition over time.
- Altered Hard Parts—after an organism is encapsulated in a liquid tomb, its hard parts may undergo alteration while its soft parts decay. The organic hard part material is replaced by minerals that have dissolved in percolating groundwater. These minerals fill pores and cavities in the specimen through chemical precipitation. If the replacement mineral is silica, the replacement process is called *petrification*. Petrification produces a replica of the original specimen that is amazingly similar—even at a microscopic level.
- Soft Parts—under certain processes, the soft parts of an organism are preserved over time. One such process is sudden freezing coupled with isolation from bacteria. Another is entombing with tree sap, such as the amber examples shown here.

Fossils in amber
courtesy Adrian Pingstone, Wikimedia Commons (in the public domain)

- Mold—this common fossil type is created when the organism decays, leaving an impression (or *mold*) in the surrounding tomb material that preserves the original shape. If the impression fills with a mineral deposit over time, the fossil is labeled a *cast*.
- Carbonization—when buried plants are heated, they expel gasses, leaving behind mostly carbon. This process results

Anthracite Coal
courtesy U.S. Geological Survey, Mineral Information Institute, Wikimedia Commons (in the public domain)

in the formation of coal, as discussed in Chapter 3. The type of buried plant can often be recognized within a coal seam. If animal bodies are compressed by the weight of sediments above them to form coal, only a thin carbon film results. As geologic processes apply pressure to biotic material over time, under suitable conditions it is transformed successively into peat (discussed in Chapter 3), lignite (brown coal), sub-bituminous coal, bituminous coal, steam coal, anthracite coal (shown here, harder and glossier than bituminous coal), and graphite.

- Trace Fossils—fossilized body waste material (*coprolite*) and stomach contents are found occasionally. The undigested food provides evidence of the animal's eating habits. Some animals swallow rocks to aid digestion; hence these smooth, rounded *gastroliths* are also sometimes found.
- Ephemeral Markings—if rapidly buried and filled with sediment, animal tracks can be fossilized, as can be worm burrows and even raindrop impressions.

Fossil record distribution

The distribution of fossils that have been found around the globe is illustrated in Figure 21 below.

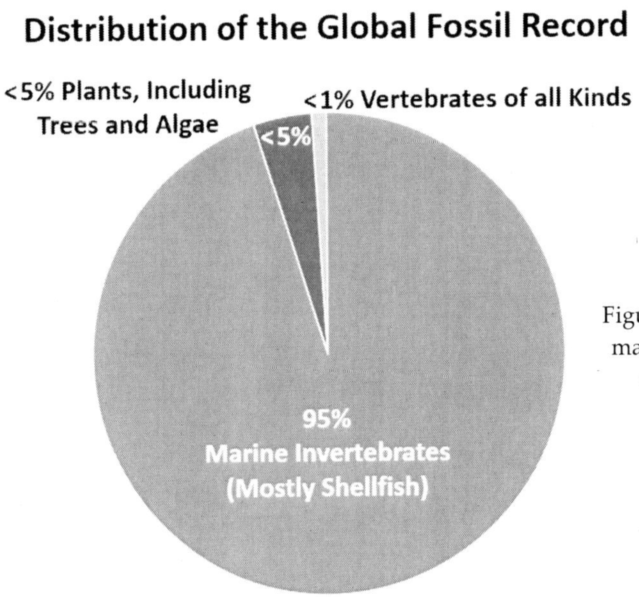

Figure 21—the vast majority of fossils are shellfish

The following paragraph from Chapter 3 explains how the catastrophic global Flood created the fossil record—

The first 150 days of the Flood caused the eventual fossilization of trillions of marine (sea dwelling) organisms and the destruction of all terrestrial (land-based) and flying creatures. First to be impacted were those organisms dwelling on the bottom of the ocean, especially near the erupting fountains. Then organisms within the shallow continental shelf areas were overcome as the initial deadly waves and underwater surges inflicted their tremendous forces. When the huge waves broke over the coastline, anything dwelling there was smothered and eventually buried in the loose sediments that were transported by the raging sea. As the pulsing walls of water climbed ever higher and farther over the supercontinent, all life within the low-lying areas, and finally the upland regions, died a horrific death.

This description actually predicts the distribution of the global fossil record shown in Figure 21. The lower sediment layers (first to be laid down on the supercontinent) were loaded with sea-dwelling animals trapped in the violent ocean currents that were generated by the breaking open of the fountains of the great deep. Those creatures with hard outer shells (such as shellfish) had the greatest fossilization potential. Other organisms submerged in the sediment-packed waters would be ground to powder by the powerful undersea mudflows. On the other hand, the upper sediment layers (last to be laid down on the supercontinent) would increasingly include terrestrial (land-based) life along with a wide variety of other fossils. However, as the Flood waters receded, with the help of a wind, the uppermost sediment layers were completely eroded, destroying any fossils that had been laid down in them.

Typically, terrestrial creatures have very low fossilization potential. This is because, once drowned, they bloat and then float, becoming easy prey for marine scavengers. A modern example of this phenomenon occurred when the *RMS Titanic* collided with an iceberg in the Atlantic Ocean and sank in 1912. 1,517 passengers and crew members were carried to a watery grave in that tragedy. Rescue efforts located some floating bodies, but the majority were trapped underwater. When the wreckage was explored in 1985—73 years later—not a single human body or body part was found. Scavengers had devoured the remains of all human life. Even cotton and wool clothing were gone!

Because of the low fossilization potential of terrestrial creatures, fossils of these animals were laid down mostly during the Ice Age that followed the Flood. The Ice Age was strictly a land-based catastrophe that had the ability to bury and rapidly freeze animal life in land-sourced sedimentary deposits. In such sediments, human fossils have been found.

Fossil locations

As discussed in Chapter 3, the Flood deposited layers of sedimentary rocks over huge areas. Some of the strata are hundreds of feet thick and cover hundreds of thousands of square miles. Each stratum is essentially unique and can be traced geographically by identifying its particular geologic characteristics as well as its fossil content. In fact, fossil content has become a convenient tool for identifying different strata. If the fossil content of a certain stratum in one location is known, this knowledge might help identify the same stratum in another place, even if erosion has separated the two locations.

The continuity of sedimentary strata cannot usually be observed directly because the strata are typically located underground. However, the drilling of oil wells provides a useful means for determining this continuity. As drill cuttings of the various layers, complete with their unique geologic and fossil content, are washed to the surface, they enable identification of the various strata.

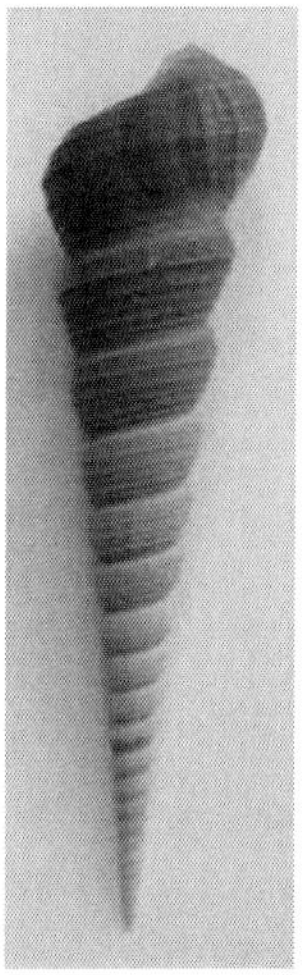

Nautiloid Shell

To illustrate the extent of the fossil record in sedimentary strata, consider the discovery of nautiloid fossils[12] in the Grand Canyon—

"About halfway down the canyon walls is a prominent reddish cliff called the Redwall Limestone, found throughout the American West. Fossils are relatively rare and unimpressive in Grand Canyon, but in 1966, within a six-foot member in the center of the Redwall, about a dozen beautiful, two-foot-long, cigar-shaped nautiloids were discovered in a remote side canyon. ... The fossil orientations are pointed in a preferred direction, indicating that moving water was

involved. Exploring adjacent side canyons, he [Dr. Steven Austin] *found other nautiloids ... Identical fossils were eventually discovered in the same layer of rock throughout Arizona, Utah, Nevada, and New Mexico---an estimated four billion sea creatures."*

The Morrison Formation was introduced in Chapter 3. Additional detail about this dinosaur fossil bonanza[13] is provided here—

"Think of the Morrison Formation of the American West, containing innumerable dinosaur remains ripped apart and quickly buried in the slurry of mud and volcanic ash. The dinosaurs appear to have lived in varied habitats, from swamp to coastal. The fossilized plant remains, however, came from a much wider range, from lowland to mountains. Mixed in with the terrestrial dinosaurs are aquatic reptiles, fish of numerous types, sharks, crocodiles, clams—clearly marine fauna. This layer covers an extensive geographic region."

The Morrison Formation once covered an even larger area than that discussed in Chapter 3. It probably covered all of the American West, including the Grand Canyon, before erosion brought it to its current size of 600,000 square miles.

In summary, fossils from all habitats are found in geologic strata that spread over vast areas. The fossils of land-dwelling creatures reside in the upper strata and typically overlie fossils from ocean-based environments.

Highlights of the Fossil Record

Referring to Figure 9 on page 181, the oldest rocks in the geologic record are assigned to the Hadean eon. They are metamorphic (formed by extreme heat and pressure) and igneous (solidified magma) rocks which have no fossil content.

Overlying the Hadean rocks are Archean eon rocks which contain fossils of *prokaryotes*—single cell organisms that lack both a membrane-bound cell nucleus and distinct cellular compartments. These tiny fossils can only be obtained by etching away the host rock with acid, and they require a high powered electron microscope to be seen. Besides obtaining life-sustaining energy from photosynthesis and organic (carbon-based) compounds, prokaryotes can also obtain energy from inorganic compounds, such as hydrogen sulfide. This flexibility allows them to thrive in harsh environments, such as the bitterly cold surface of Antarctica and the extremely hot undersea fissures of black smokers. The modern *kind* of prokaryotes includes bacteria and blue-green algae.[14, 16]

Stromatolites are also found in the Archean rocks. Layered mat structures that are secreted by prokaryotes, stromatolites are very similar to modern algae deposits. An example of stromatolites found in Glacier National Park in Montana is shown here.

Stromatolites
courtesy U.S. National Park Service, Wikimedia Commons (in the public domain)

Besides the prokaryotes and stromatolites discussed above, the Proterozoic eon rocks contain fossils of *eukaryotes*—organisms that have both a membrane-bound cell nucleus and distinct cellular compartments. The definition of a eukaryote encompasses all forms of life except the prokaryotes discussed above. However, only eukaryotes that were single celled or grouped in communities appear in the Proterozoic fossil record.[18]

The next layer of the fossil record is associated with the Paleozoic era and is known as the Cambrian Explosion. This important geologic and fossil milestone was introduced in Chapter 3. The Cambrian Explosion fossil layer contains virtually every basic marine animal type—from invertebrates to vertebrate, jawless fish. Fossils of trilobites (now extinct) are extremely plentiful, revealing eyes that are as functional as any found in animal life today. Fossils of clams, brachiopods, starfish, and cephalopods, like those shown on the next page, are also plentiful. The animals that formed these fossils are very much like their living counterparts today.

The author owns some property on top of Kanarra Mountain in southeastern Utah that is at an elevation of 9,200 feet (2804 m) above sea level. Fossils of trilobites have been found on this property.

Besides the sea life identified above, the Cambrian Explosion sediment layers also include fossils of seaweeds, worms, sponges, mollusks, corals, jellyfish, sea cucumbers, and stromatolitic reefs. About 30% of the fossils[19] are brachiopods, and 60% are trilobites (more than 10,000 types).

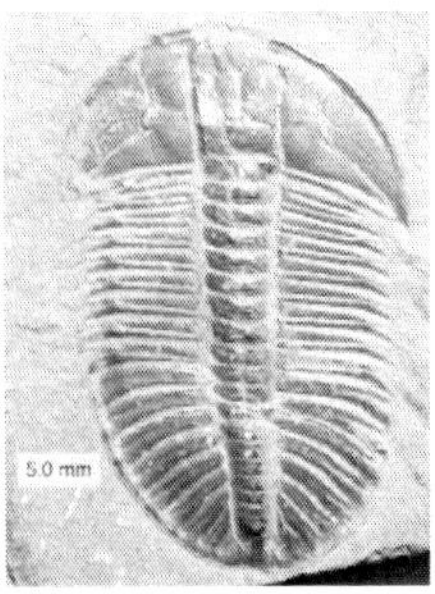

Trilobite
courtesy Mark Wilson, Wikimedia Commons (in the public domain)

Brachiopod
courtesy Mark Wilson, Wikimedia Commons (in the public domain)

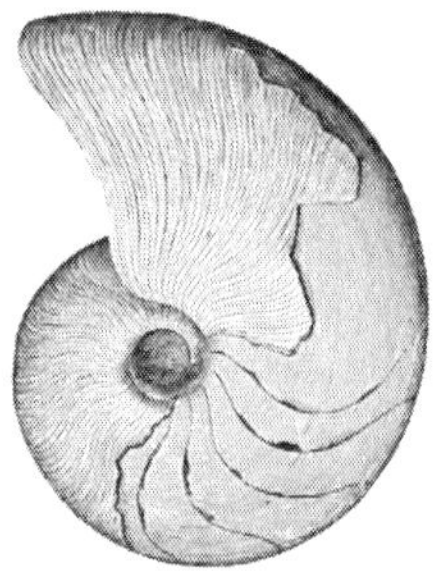

Cephalopod
(*courtesy Wikimedia Commons (in the public domain)*

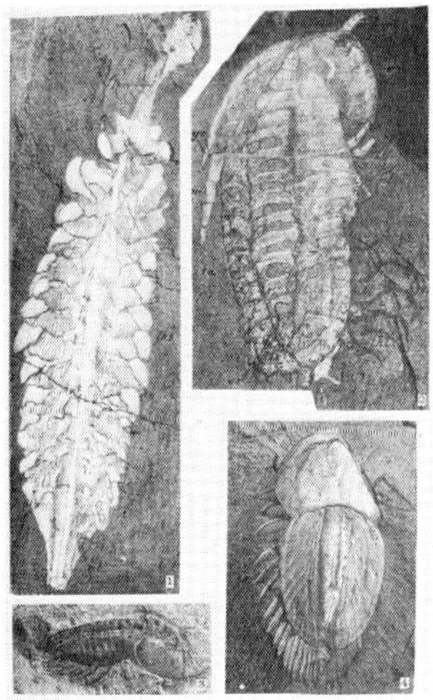

Fossils collected by Walcott, including *Opabinia* (upper left)
(in the U.S. public domain)

In 1909, Dr. Charles Walcott (the paleontologist introduced previously who proposed that the Lipalian interval of the geological clock was the period of time when ancestors of Cambrian animals evolved, but that no fossil record of them was preserved) discovered the Burgess Shale Formation, located in the Canadian Rockies of British Columbia. This formation is important to the fossil record because the extremely fine-grained shale that comprises the stratum preserved intricate, minute details of previously unknown invertebrates. It is also one of the earliest fossil beds containing soft part imprints. The formation is dated by modern geology at Middle Cambrian, which puts it more recent than the Cambrian Explosion. Thus, the previously unknown animal types that it reveals—without any trace of fossilized predecessors—can be added to those of the Cambrian Explosion, making it even more difficult for evolutionists to explain how so many different types of animals appeared suddenly without evidence of transitional forms. Some of the animals that Dr. Walcott discovered in the Burgess Shale Formation fossil record had rather bizarre anatomical features, such as the five-eyed *Opabinia* shown above.[21]

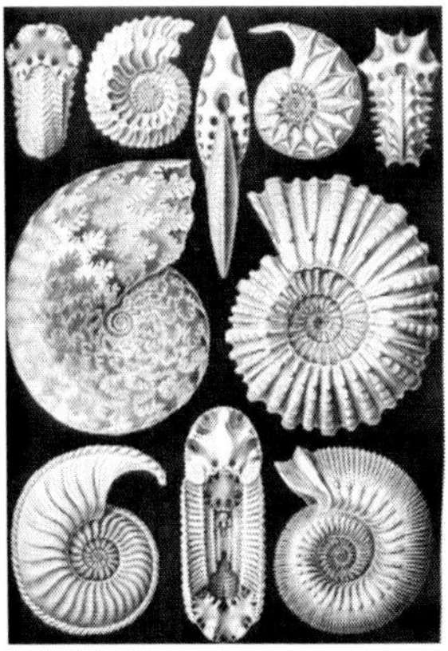

Ammonite drawings by Ernst Haeckle (1904 *Art Forms of Nature)*
(in the U.S. public domain-expired copyright)

Clams and other marine invertebrate fossils are found in many locations around the world, including the peaks of high mountains such as Mount Everest (the tallest mountain on earth) and Kanarra Mountain, where the author owns property. Ammonite fossils (some examples are shown here) are among the most common. When these invertebrates were alive, they were ocean bottom dwellers. Thus they were among the first animals to be fossilized by the Flood. Catastrophic plate tectonics buckled up ocean bottom rock strata containing such fossils to become mountain ranges when tectonic plates collided during the Division of the earth. It has been estimated that trillions of invertebrate fossils exist around the globe.

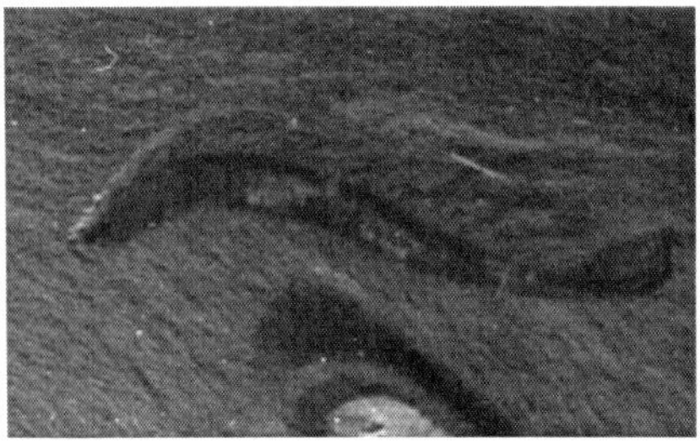

Pikaia fossil
courtesy Jstuby at the wikipedia project, Wikimedia Commons (in the public domain)

An extinct fossil known as *Pikaia* was discovered in the Burgess Shale that many evolutionists consider to be the oldest ancestor of modern vertebrates. Less than 2 inches (5 cm) long, it supposedly swam above the ocean floor much like an eel. Pikaia is of interest to evolutionists because fossil details clearly show *chordate* features—meaning that a central nerve chord, supported by a rod-like structure, runs along the full length of the body. However, the Pikaia did not have a true backbone; nor did it have a recognizable brain or even a head. It did not have teeth or a heart, either. Yet evolutionists claim that all vertebrates (including man) evolved from an animal something like this.

Interestingly, a modern fish-like animal called a lancelet (*Amphioxus*) is nearly identical to the Pikaia, and is therefore a "living fossil" that has not evolved in a half billion years (according to the Geologic timescale). Furthermore, recent fossil discoveries in the lowest Cambrian strata that comprise the Cambrian Explosion are clearly vertebrate fish. They are jawless and often have bony scales for skin, but are fish nonetheless. **How could a vertebrate fish predate the original ancestor of vertebrates?** Interestingly, representatives of these strange Cambrian Explosion fish types exist today—they are known as armored fish.

In his book *Science and Earth History—The Evolution/Creation Controversy*, evolutionist author Arthur Strahler endorses Pikaia as an ancestor of vertebrates, but he also admits[22] there is an—

"absence of the traditional fossils in the gaps between each group of fishes and its ancestor ..."

After this admission, Strahler adds a surprising note[22] (highlights are the author's, not Strahler's)—

"This is one count in the creationists' charge that can only evoke in unison from the paleontologists a plea of ***nolo contendere*** [no contest]."

Regarding the evolutionary transformation of fish into tetrapods (the first four-limbed vertebrates), the lobe-finned fish (*Coelacanth*) was the most popular choice of a tetrapod ancestor for nearly a century. Then in 1938, a live lobe-finned fish was found among the haul of a fishing boat near Madagascar. At last, evolutionists could study a real missing link!

However, results of the study dashed evolutionary hopes and expectations. No feature of the Coelacanth was adapted to life on land. The fish's heart, brain, and gut were vastly different from what had been expected. Furthermore, studies later proved that the fish lives in deep water—not shallow water where it was supposed to have acquired the facilities to move onshore. Also, the strong fins along the bottom of the fish are used to make difficult swimming maneuvers possible in the deep sea—not to support the fish's weight. Since 1938, Coelacanth fish have been discovered in other areas around South Africa as well as the West Indian Ocean. Fossils of Coelacanth fish have been found in rocks dated 400 million years old, yet the modern fish is essentially the same! Thus, stasis (no evolutionary change) has supposedly gone on for 400 million years.

The current darling of evolutionists to replace Coelacanth as the transformation link for tetrapods was discovered in 2004 on Ellesmere Island, part of the Canadian Territory of Nunavut and one of the Queen Elizabeth Islands. It is called *Tiktaalik*. Dated at 375 million years, this fossil fish had traits of both fish and amphibians. For example, it had an amphibian-like flat head with eyes on the top, not on the sides. It also had an unusual arrangement of bones within its front fins, but no such bones were in its rear fins. It was definitely a fish, for it had gills, scales, and fins and lived in water. Tiktaalik was about 10 feet (3 m) long and is thought to have had a lung-like organ.[23]

When Tiktaalik was first discovered, the news media promoted the discovery as exciting, major news for evolution. Tiktaalik was likened to mudskippers, crocodilefish, and alligator gars—all living creatures today. (Incidentally, fossil gar fish are nearly identical to modern gars.) However, many evolution experts saw weaknesses in Tiktaalik as a transitional form, and prefer other candidates.

Then in 2010, Uppsala University (Sweden's oldest university) released a press dispatch[24] that stated—

"But there is another potential source of information about the earliest tetrapods: the fossilized footprints they left behind. In the new study a Polish-Swedish team describe a rich and securely dated footprint locality from Zachelmie Quarry in Poland that pushes back the origin of tetrapods a full 18 million years beyond the earliest skeletal evidence and forces a dramatic reassessment of the transition from water to land." And further—*"These results force us to reconsider our whole picture of the transition from fish to land animals, says Per Ahlberg of Uppsala University, one of the two leaders of the study."*

One would expect this discovery of fossilized tetrapod footprints, dated 18 million years earlier than Tiktaalik, to cause evolutionists to reject Tiktaalik as a missing link between fish and amphibians. However, this did not happen. In the spirit of Lewontin's "unsubstantiated just-so stories," the press release quoted above also declared (highlights are the author's)—

"This means that not only tetrapods but also elpistostegids [Tiktaalik] *originated much earlier than we thought,* ***because the position of elpistostegids as evolutionary precursors of tetrapods is not in doubt, and so they must have existed at least as long,*** *says Per Ahlberg."*

Consequently, Tiktaalik is still touted as a transitional form between fish and amphibians.

The author marvels how modern evolutionary science can repeatedly get away with rewriting fossil history "because it must be so" without incurring the ire of more disciplined scientists. The author reminds the reader of the words of Dr. Michael Ruse,[25] who is widely considered the leading spokesman for evolution today—

"Evolution is promoted by its practitioners as more than mere science. Evolution is promulgated as an ideology, a secular religion—a full-fledged alternative to Christianity, with meaning and morality. I am an ardent evolutionist and an ex-Christian, but I must admit that in this

one complaint ... the literalists are absolutely right. Evolution is a religion. This was true of evolution in the beginning, and it is true of evolution still today."

Adopting evolution as a religion should not condone the use of unsubstantiated just-so stories as the means for developing its doctrine!

One evolutionist—the late Dr. Barbara Stahl—was intellectually honest enough to publish a 544 page book in 1974, titled *Vertebrate History: Problems in Evolution,* that documented the numerous problems involved with evolution of the vertebrates. A second edition was published in 1985. Of course, these books were an embarrassment to most evolutionists, who tried to downplay Dr. Stahl's research. Both books are now out of print and are conveniently considered "out of date." Nevertheless, Dr. Stahl wrote this[26] about the evolution of reptiles—

"It is easier to understand the stages by which the reptiles evolved temporal fenestra [certain features of the skull] and other distinguishing skeletal characters than to imagine the steps that led to the development of the 'land egg' ... Unraveling the tangled history of [early reptile] origins has been a slow process because of the limited amount of useful fossil material and the difficulty of studying it."

A more recent book, *Integrated Principles of Zoology,*[27] written by Cleveland Hickman and two co-authors in 1997, agreed with Dr. Stahl—

"Understanding the 300-million-year old history of reptile life on earth has been complicated by ... large gaps in the fossil record."

Perhaps one of the most "absurd constructs" to which Dr. Lewontin referred is the story of marine reptiles. Evolutionists claim that, after reptiles evolved into land based creatures, some of them returned to the sea, such as the ancestors of the *Ichthyosaur*. This twist to evolution would require Darwinian evolution to operate in reverse!

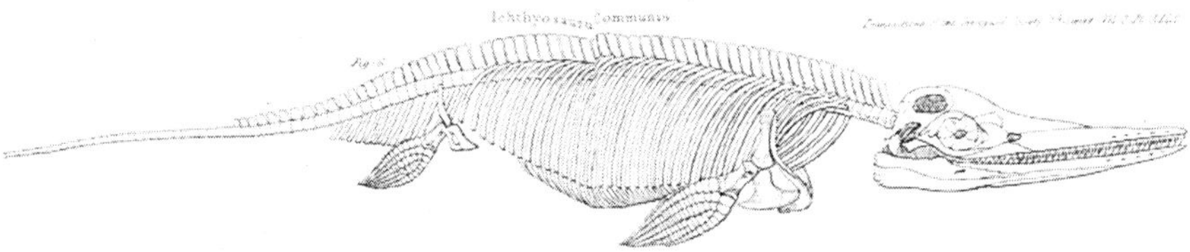

Skeletal anatomy of *Ichthyosaurus* from an 1824 paper by Conybeare
(in the U.S. public domain-expired copyright)

Evolutionists argue that "survival of the fittest" by natural selection was not relevant when the evolving ancestors of reptiles (misfit mutants) first went onto land—because there was no competition for survival there (they were the first land-based creatures). In order for Darwinian evolution to operate in reverse, when the evolving ancestors of marine reptiles (also misfit mutants) went back into the ocean, "survival of the fittest" by natural selection could not have been relevant there either—or else the functionally inferior mutants would quickly be attacked by predators. How absurd to make this assumption! By then, the oceans would have been teeming with predators and myriads of other competitors, so natural selection would have been in full swing!

To compound this absurdity, evolutionists also claim that development of the ancestors of marine reptiles paralleled development of the ancestors of modern dolphins and whales—"which they resembled in a case of *convergent evolution*." [28]

"Convergent evolution describes the independent evolution of similar features in species of different lineages. Convergent evolution

creates analogous structures that have similar form or function, but were not present in the last common ancestor of those groups." For example, *"Flying insects, birds, and bats have all evolved the capacity of flight independently. They have 'converged' on this useful trait."* [29]

Incidentally, the extinct marine reptile ichthyosaur gave birth to live young rather than laying eggs (nearly all other female reptiles lay eggs). Some female ichthyosaur fossils have been found that were actually preserved in the process of giving live birth. In order for these fossils to have been made, burial of the mother and *neonate* (child) must have happened extremely fast, as would have occurred during the Flood. Live birth was not employed by any of the proposed ancestors of ichthyosaur. In the fossil record, ichthyosaurs appear abruptly, fully formed and functional, without evidence of any ancestral links. According to Dr. Edwin Colbert, Curator Emeritus for the American Museum of Natural History and Professor Emeritus at Columbia University—

"The basic problem of ichthyosaur relationships is that no conclusive evidence can be found linking these reptiles with any other reptilian order." [30]

Plesiosaur skeleton
courtesy Wikimedia Commons (in the public domain - expired copyright)

Other reptiles that have no fossil evidence of ancestral links are the ocean-based *Plesiosaur* and the flying *Pterodactyloids* and *Rhamphorhynchoids.* Evolutionists believe plesiosaur descended from land reptiles. Up to 50 feet (15 m) long, this amazing reptile had large paddles and a long neck, and is unlike any other reptile.

Pterodactylus antiquus cast
courtesy Daderot, Wikimedia Commons (in the public domain)

Evolutionists have determined that the flying reptiles must have descended from a walking dinosaur, although they form a group that is unlike any other reptile. They are also quite different from each other. The pterosaurs, like the one shown here, had a long, bony crest on top of the skull and a wing span as great as 54 feet (16.5 m), while the rhamphorhynchoids typically had teeth and long tails with a bulge at the end. Like the ichthyosaur, plesiosaur, rhamphorhyncus, and pterodactylus appear abruptly in the fossil record fully developed.

One reptile that has a high fossilization potential is the turtle. Because of their hard shells, which resist decay, turtle fossils are easy to identify, even if shattered. Turtles are unlike any other animal. Dr. Edwin Colbert wrote the following[31] about the turtle fossil record—

"The first turtles made their appearance by the late part of the Triassic period, by which time they were far advanced along the lines of adaptive radiation typical of modern turtles."

Evolutionists have suggested transitional ancestors, but of course none have been found. The authors of *The Fossil Record* wrote[32] the following about the turtle fossil record—

"This should be an excellent test case for evolution. The signature features are so easily fossilized, so recognizable, and so different from those of any other earlier creature that surely transitional forms would be discovered if they ever existed. They are, however, nowhere to be found. No reptile even possesses a feature that could be modified to become either carapace [the upper shell] or plastron [the lower shell]. A spreading and hardening of the ribs will not do. They are completely novel features."

Snakes provide another example of a fossil record that has no transitional ancestors, even though there are thousands of snake and lizard fossils that have been discovered around the world. Snakes are thought to have evolved from lizards. According to Michael Benton, author of the textbook[33] *Vertebrate Palaeontology* for graduate students—

"The snakes ... are believed to have arisen from 'lizard' ancestors, but the nature of those ancestors is a mystery."

Dr. Edwin Colbert wrote the following in his textbook (highlights are the author's, not Colbert's)—

"Unfortunately, the fossil history of the snakes is very fragmentary, ***so that it is necessary to infer much of their evolution from the comparative anatomy of modern forms****."* [34]

In response to this statement of a highly respected paleontologist, the authors of *The Fossil Record* wrote[32]—

"Once again, the assumption of evolution negates the need for evidence, and ignores the lack of evidence supporting its assertions."

Perhaps the most fascinating part of the fossil record involves dinosaurs—creatures that intrigue nearly everyone, young and old. Relatively abundant dinosaur fossils have been found on every continent, revealing a large variety of these interesting reptiles. Although some were very large, most dinosaurs were similar in size to common animals of today, such as large birds, dogs, and horses.

Paleontologists define dinosaurs mostly by the shape of their skull and type of hip. All dinosaurs had *fenestrae* (natural openings in the skull), located either in front of or behind the eyes. These openings reduced the weight of the skull. Every dinosaur also had one of two types of hips, which are used to classify them into two major orders. *Saurischia* dinosaurs had what is called a "lizard hip," even though it did not resemble a true modern lizard hip. All carnivorous (meat eating) dinosaurs, and some herbivorous (plant eating) dinosaurs, were saurischians. *Ornithischia* dinosaurs had what is called a "bird hip," even though it did not resemble a true modern bird hip. Bird-hipped dinosaurs had beaks and were herbivores. Both bird-hipped and lizard-hipped dinosaurs walked with their legs directly beneath their bodies, as mammals do today. Ironically, modern reptiles with legs do not walk like this. Their legs extend outward from their bodies (picture lizards and crocodiles) so they can start and stop running very quickly.

Saurischian dinosaurs supposedly evolved first, in great variety. Ornithischian dinosaurs followed, but from what did they evolve? According[35] to Dr. Barbara Stahl—

"Paleontologists have virtually no clues to the origin of the ornithischian (bird-hipped) dinosaurs. They appear in the fossil record in increasing numbers after the middle of the Jurassic period, already structurally distinct from the saurischians (lizard-hipped dinosaurs), without any telltale traits which might link them to a particular group of earlier archosaurs."

Reading to this point, the reader has undoubtedly realized that the fossil record does not support Darwinian evolution for three reasons—it does not contain fossils of transitional forms; the fossils it does contain are fully developed when they first appear; and these fossils remain essentially the same until they no longer appear.

The fact that dinosaur fossils are found all around the world and from pole to pole is explained in Chapter 3. The pre-Flood earth had a single supercontinent graced with a subtropical climate everywhere, allowing dinosaurs to thrive over all its land surfaces and in the ocean. Catastrophic plate tectonics broke the supercontinent into fragments, which then moved apart to form the continents which exist today.

When the Flood did its most destructive work during the first 150 days, it violently destroyed all terrestrial animal life by burying the animals in sedimentary slurries and then spreading the slurries over the supercontinent. These mud slurries contained the remains of marine plants and animals, including sea-based dinosaurs, as well as terrestrial plants and animals—all mixed together. Many bodies were broken and dismembered. This is the reason that the Morrison Formation, described previously, and the Dinosaur National Monument Quarry (which is part of the Morrison Formation) have many more fossils of clams and other aquatic animals than fossils of land-based dinosaurs. The Flood also explains why dinosaur skeletons are piled on top of one another in this Formation, as well as globally.

Because the Flood created the fossil record, no fossil deposit anywhere on earth provides a snapshot of the environment in which the fossilized plants and creatures actually lived when the Flood waters overtook them.

Fossils of dinosaurs, pterosaurs, birds, and mammals often exhibit evidence of drowning. Such evidence includes wide-open mouths, heads thrown back as if the animals were gasping for air, and tails that arch toward the head. Many fossils of clams are found in self-protect mode, with both halves tightly closed. Thus fossilized animals are best understood as having been buried alive in a watery grave along with specimens from several different environments picked up on the way by the rapidly moving slurries.

Breakthrough Science Regarding the Dinosaur Fossil Record

This book has explained why the dating methods used by modern science to determine the age of fossils are unreliable. So when did the dinosaurs really live?

The answer to this question may surprise the reader. The story behind the answer begins in 1908 when fossil collector

Charles Sternberg and his three sons found the mummy of a duckbilled dinosaur in the State of Wyoming. They found the mummy lying on its back with its head pulled under its body and its right arm stretched into the air. The skin of the chest and abdominal region was pulled into the body cavity. Unfortunately, the tail, hind feet, and hind portion of the pelvis had eroded away. Figure 22 below shows this mummy on display at the American Museum of Natural History in New York. Figure 23 (see the next page) provides an abdominal view, and Figure 24 (also next page) provides an impression of the abdominal skin. The mummy was officially determined to be an *Edmontosaurus (or Trachodon) annectens* that lived 66 million years ago.

Figure 22—*Trachodon* mummy found near
Lusk, Wyoming in 1908
courtesy Wikimedia Commons (in the public domain - expired copyright)

How could an animal 66 million years old be a mummy today? Why has it not disintegrated?

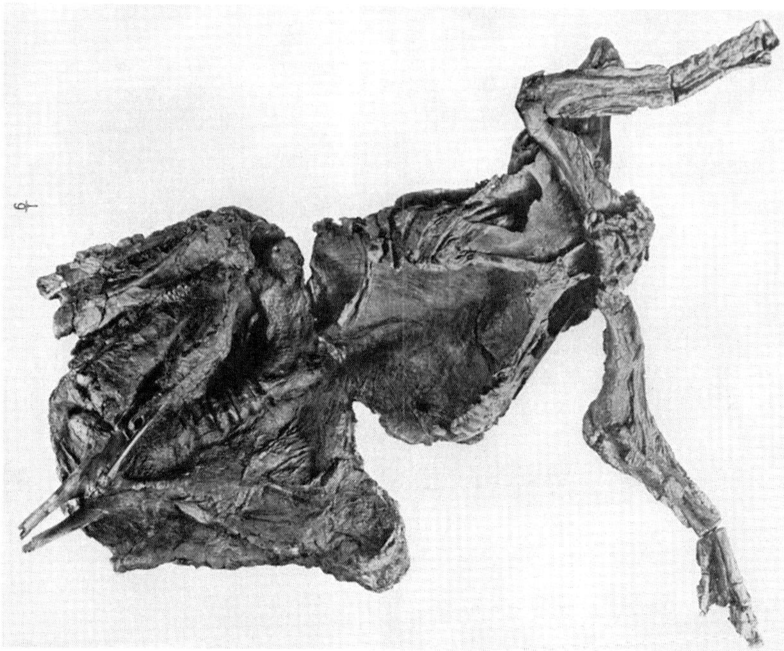

Figure 23—Abdominal view of the *Trachodon* mummy

courtesy Wikimedia Commons (in the public domain - expired copyright)

Figure 24—Abdominal skin impression from the *Trachodon* mummy

courtesy Wikimedia Commons (in the public domain - expired copyright)

The story continues with another discovery (this time accidental) in 2004 by Dr. Mary Schweitzer when the thigh bone of a *Tyrannosaurus rex* (*T. rex*) was broken in half for transportation purposes. Inside a sample of the fractured bone, Dr. Schweitzer found *original* soft tissue! A paper that she wrote, published[36] in the journal *Science* in March 2005, contained the following abstract—

"Soft tissues are preserved within hindlimb elements of Tyrannosaurus rex (Museum of the Rockies specimen 1125). Removal of the mineral phase reveals transparent, flexible, hollow blood vessels containing small round microstructures that can be expressed from the vessels into solution. Some regions of the demineralized bone matrix are highly fibrous, and the matrix possesses elasticity and resilience. Three populations of microstructures have cell-like morphology. Thus, some dinosaurian soft tissues may retain some of their original flexibility, elasticity, and resilience."

Not only was the soft tissue stretchy and pliable, its hollow blood vessels contained actual blood cells!

How could a fossil dated 68 million years old still contain soft tissues and red blood cells?

Even under ideal preservation conditions, organic tissues are known to decay rapidly, especially if water is present. When discovered, the T. rex thigh bone was entombed in a porous sandstone in Montana—certain to have been penetrated by groundwater. So how could the bone's organic material have escaped decomposition for so many millions of years?

Fragile proteins and enzymes were also present in the T. rex sample that Dr. Schweitzer studied, indicating that this specimen was much younger than its assigned age. The sample even had *medullary bone* (a thin layer of highly vascular, estrogen-linked reproductive bone tissue that is found in modern female birds only during ovulation), indicating that the T. rex was a female who died during her egg laying cycle.[37]

An article[38] by Helen Fields that was published in the *Smithsonian Magazine* in May 2006 added some additional detail (note: B. rex was the nickname given to the specimen Dr. Schweitzer studied)—

"At an October 2005 paleontology conference, Dr. Schweitzer presented preliminary evidence that she had detected real dinosaur proteins in her specimens. ... Further discoveries in the past year have shown that the discovery of soft tissue in B. rex wasn't just a fluke. Schweitzer and Wittmeyer have now found probable blood vessels, bone-building cells and connective tissue in another T. rex, in a theropod from Argentina, and in a 300,000-year-old woolly mammoth fossil."

Of course, publication of Dr. Schweitzer's paper in 2005 brought immediate criticism from evolutionists, as well as attempts to refute the implications of her work. Many called for replication of her findings and more proof of what she had found, refusing to accept the initial results. Their demands were answered.[39] In the May 1, 2009 edition of the journal *Science,* Robert Service wrote the following "News of the Week" teaser—

"A controversial finding that protein fragments can be recovered from dinosaur fossils has been replicated for the first time. Two years ago, a team of paleontologists stunned the community when they reported discovering intact protein fragments in a fossil from a Tyrannosaurus rex *that died 68 million years ago. The claim has remained contentious, because proteins in tissue normally degrade quickly after an animal dies. On page 626 of this week's issue of* Science, *however, the team reports finding an even larger number of protein fragments from an 80-million-year-old fossil from a duck-billed dinosaur, or hadrosaur, known as* Brachylophosaurus canadensis.*"*

In order to address the criticism and contention that she had faced after documenting the results of her first discovery, Dr.

Schweitzer recruited fifteen additional collaborators to work with her on her second[40] paper, which included the following in the abstract—

"Molecular preservation in non-avian dinosaurs is controversial. We present multiple lines of evidence that endogenous proteinaceous material is preserved in bone fragments and soft tissues from an 80-million-year-old Campanian hadrosaur, Brachylophosaurus canadensis *[Museum of the Rockies (MOR) 2598]. … These data complement earlier results from* Tyrannosaurus rex *(MOR 1125) and confirm that molecular preservation in Cretaceous dinosaurs is not a unique event."*

Not surprisingly, once Dr. Schweitzer published her results, other scientists came forward with similar findings, some of which had not been reported previously because of fear of backlash like Dr. Schweitzer encountered.

The table below summarizes 39 published reports of original soft tissue fossils.[41]

Articles Published in Peer-Reviewed Journals

Publication Date	Brief Description	Evolutionary Age in Millions of Years	Publication
5/30/1977	Catfish fatty fin in Green River	50	H.P. Buchheim & R.C. Surdam, *Geology*, 5: 198
6/14/1992	Osteocalcin in a seismosaur bone	150	Muyzer, G. et al, *Geology*, 20: 871-874
9/25/1992	DNA in amber	30	Morell, V. et al, *Science*, 257: 1860

Publication Date	Brief Description	Evolutionary Age in Millions of Years	Publication
6/16/1994	Unaltered amino acids in amber insects	130	Bada, J.L. et al, *Geochemica et Cosmochemica Acta,* 58 (14): 3131-3135
6/16/1994	Dinosaur DNA from hadrosaur bone	65	Woodward, S.R. et al, *Science,* 266 (5188): 1229-1232
5/19/1995	Live bacteria spores from amber	25-40	Cano, R.J. and M.K. Borucki, *Science,* 268 (5213): 1060-1064
6/10/1997	Hemoglobin fragments in *T. rex* bone	67	Schweitzer, M. et al, *PNAS,* 94 (12): 6291-6296
6/2/1999	Live bacteria from halite deposit	250	Vreeland, R.H. et al, American Society for Microbiology, 99th General Meeting, June 2, 1999, Chicago
6/21/1999	Live bacteria from separate rock salts	250	Stan-Lotter, H. et al, *Microbiology,* 145 (12): 3565-3574
6/21/1999	Ichthyosaur skin	190	Linghan-Soliar, T. et al, *Proc. Royal Soc. B,* 266 (1436): 2367-2373
6/21/1999	Keratin in Madagascar Cretaceous bird	65	Schweitzer, M.H. et al, *J. Vert. Paleo,* 19 (4): 712-722

Publication Date	Brief Description	Evolutionary Age in Millions of Years	Publication
9/1/2001	*T. rex* collagen SEM scans	65	Armitage, M., *Creation Research Society Quarterly*, 38 (2) 61-66
6/26/2004	Live (non-spore) bacteria in amber	120	Greenblatt, C.L. et al, *Microbial Ecology*, 48 (1): 120-127
3/24/2005	*T. rex* soft tissue	68	Schweitzer, M. et al, *Science*, 307: 1952-1955
7/25/2006	Soft frog, intact	10	McNamara, M. et al, *Geology*, 34: 641-644
6/30/2007	*T. rex* collagen	68	Schweitzer, M. et al, *Science*, 316: 277-280
8/1/2007	Bloody frog bone marrow	10	McNamara, M. et al, *Geology*, 34 (8): 641-644
4/7/2008	*Psittacosaurus* skin	125	Linghan-Soliar, T. et al, *Proc. Royal Soc. B*, 275: 775-780
7/8/2008	Feather melanocytes	100	Vinther, J. et al, *Biology Letters*, 4: 522-525
4/30/2009	Hadrosaur blood vessels	80	Schweitzer, M. et al, *Science*, 324 (5927): 626-631
5/19/2009	Primate "Ida" soft body outline	40	Franzen, J.L. et al, *PLoS ONE*, May 2009 4 (5): e5723

Publication Date	Brief Description	Evolutionary Age in Millions of Years	Publication
7/1/2009	Hadrosaur skin cell structures	66	Manning, P. et al, *Proc. Royal Soc. B*, 276: 3429-3437
8/26/2009	Purple Messel feather nanostructure	40	Vinther, J. et al, *Biology Letters*, 6 (1): 128-131
10/2/2009	Fungal chitin ubiquitous in Permo-triassic	250	Jin, Y.G. et al, *Science*, 289 (5478): 432-436
8/18/2009	Squid ink	150	Whilby, P.R. et al, *Geology Today*, 24 (3): 95-98
11/5/2009	Salamander muscle, whole	18	McNamara, M. et al, *Proc. Royal Soc. B*, 277 (1680): 423-427
2/25/2010	*Sinosauropteryx* melanosomes	125	Zhang, F. et al, *Nature*, 463: 1075-1078
5/14/2010	Mammal hair in amber	100	Vullo, R., *Naturwissenschaften*, 97 (7): 683-687
3/10/2010	*Psittacosaurus* skin color	125	Linghan-Soliar, T.G. and Plodowski, *Naturwissenschaften*, 97: 479-486 (Note: same sample analyzed in *Proc. Royal Soc. B*, 275: 775-780)

Publication Date	Brief Description	Evolutionary Age in Millions of Years	Publication
5/18/2010	*Archaeopteryx* original tissue	150	Bergmann, U., *PNAS*, 107 (20): 9060-9065
8/9/2010	Mosasaur blood, retina	65-68	Lindgren, J., *PLoS ONE*, 5 (8): e11998
11/12/2010	Penguin feathers	36	Clarke, J.A. et al, *Science*, 330: 954-957
11/18/2010	Shrimp shell and muscle	360	Feldman, R.M. and C.E. Schweitzer, *J. Crustacean Biology*, 30 (4): 629-635
2/7/2011	Chitin and chitin-associated protein	417	Cody, G.D. et al, *Geology*, 39 (3): 255-258
4/1/2011	^{14}C date of mosasaur (24,600 Yrs)	70	Lindgren, J. et al, *PLoS ONE*, 6 (4): e19445
3/23/2011	Lizard tail skin, Green River	40	Edwards, N.P. et al, *Proc. Royal Soc. B*, online
6/8/2011	Type I Collagen, *T. rex* and hadrosaur	68	San Antonio, J.D. et al, *PLoS ONE*, 6 (6): e20381
6/3/2011	Bird feather pigment	120	Wogelius, R.A. et al, *Science*, online
7/2013	Soft sheets of fibrillar bone	67	Armitage, M. *Acta Histochemica*, 115 (6): 603-608

Another amazing discovery was summarized in an online article[42] written by Murray Wardrop, posted to *The Telegraph* on August 19, 2009—

Scientists draw squid using its 150 million-year-old fossilised ink

> Scientists unearthed a fossilised squid which was so well preserved that they were able to draw a picture of it using its own 150 million-year-old ink. Paleontologists discovered the remains of the creature, called a *Belemnotheutis antiquus*, during a dig at a Victorian excavation in Trowbridge, Wilts. They cracked open what appeared to be an ordinary looking rock only to find the one-inch-long black ink sac inside. After realising what they had stumbled across, they took out a small sample of the black substance and ground it up with an ammonia solution. Remarkably, the ink they created was good enough to allow them to draw the squid-like animal and write its Latin name.

A similar article[43] by David Derbyshire gives more startling information—

"... the black ink was of exactly the same structure as that of today's version."

Thus, no evolutionary change to the ink had occurred—after, supposedly, 150 million years!

Even more amazing than the discoveries discussed above was the discovery in Canada of a nearly perfect dinosaur mummy[44] that has its original skin, soft tissues, and even stomach contents intact!

"For the first time in human history a dinosaur fossil is presenting us with the opportunity to see what a living dinosaur looked like. In 2001 a field research team unearthed the world's most complete dinosaur fossil, a Brachylophosaurus nicknamed Leonardo. This incredible dinosaur has all of its skin covering its body, missing only a bit on the

head and tail. Leonardo has more soft tissue preserved than any other dinosaur fossil, and may even have some of its internal organs. The subject of intensive scientific study, the world is about to discover more about dinosaurs than ever before! More information can be found at www.dinosaurmummy.org, along with the story of his discovery."

The author recommends that the reader visit the website www.dinosaurmummy.org to watch a January 2003 video of a paleontologist seeing Leonardo for the first time.

What do these discoveries of dinosaur mummies, soft tissues, blood cells, and even proteins say about the fossil record? According to mainstream scientists, the message is merely that these fossils received extraordinarily superior preservation. Dr. Mary Schweitzer is one such scientist. In January 2014, a paper she[45] and seven collaborators wrote appeared in the *Proceedings of the Royal Society B* [Biological Sciences]. An announcement[46] of this article, made by North Carolina State University (where Dr. Schweitzer works) on November 26, 2013, stated the following—

"Iron may play a role in preserving ancient tissues within dinosaur fossils, but also may hide them from detection. ... Schweitzer's latest research shows that the presence of hemoglobin—the iron-containing molecule that transports oxygen in red blood cells—may be the key to both preserving and concealing original ancient proteins within fossils. ... Schweitzer and her team noticed that iron particles are intimately associated with the soft tissues preserved in dinosaurs. But when they chelated—or removed the iron from—soft tissues taken from a T. rex *and a* Brachyolophosaurus, *the chelated tissues reacted much more strongly to antibodies that detect the presence of protein, suggesting that the iron may be masking their presence in these preserved tissues. They then tested the preservation hypothesis by using blood vessels and cells taken from modern ostrich bone. They soaked some of these vessels in hemoglobin taken from red blood cells, while placing other vessels in water. Two years later, the hemoglobin-treated soft vessels remained intact, while those soaked in water degraded in less than a week. 'We*

know that iron is always present in large quantities when we find well-preserved fossils, and we have found original vascular tissues within the bones of these animals, which would be a very hemoglobin-rich environment after they died,' Schweitzer says. 'We also know that iron hinders just about every technique we have to detect proteins. So iron looks like it may be both the mechanism for preservation and the reason why we've had problems finding and analyzing proteins that are preserved.'"

Thus Dr. Schweitzer believes she has found the source of superior preservation that led to her discovery of soft tissues in a T. rex thigh bone. The mainstream evolution community undoubtedly sighed with relief when Dr. Schweitzer's article was published, for it staved off the other logical explanation that could be given for the discoveries of dinosaur mummies, soft tissues, blood cells, and proteins covered above—**that the dinosaur fossils are *not* millions of years old; rather, they resulted from the Great Flood about 4,500 years ago.**

On December 11, 2013, a response[47] to Dr. Schweitzer's paper by science writer Brian Thomas was released that analyzed the work Dr. Schweitzer and her colleagues did. In essence, Thomas pointed out that—

1. In order to test the iron preservation theory, ostrich vessels were first incubated in a concentrated solution of red blood cells. This procedure effectively extracted and purified iron from blood. Such a process does not happen naturally when an animal dies; therefore, testing the theory with an artificially created blood solution rendered the test results invalid.

2. Many papers identified in the Table above, which describe the discovery of soft tissues, do not provide evidence of iron in the immediate vicinity of the soft tissues.

3. The control sample—ostrich vessels continuously soaked in water—did not represent natural burial and fossilization conditions; therefore, an artificially high decay rate difference resulted.

4. The test showed that iron inhibits microbes, but it did not show that it inhibits or even slows down the oxidation and hydrolysis reactions that transform soft tissues into particles of dust.

5. While the invalid test results demonstrated how iron increases resistance to microbes for two years, it did not show that this happens for millions of years. Thus the results fall far short of the authors' claim that iron enables soft tissues to persist for tens of millions of years.

6. However, the research may explain how soft tissues have resisted disintegration for a certain amount of time—perhaps thousands of years.

The reader may be interested to assess whether or not the following verses in the book of Job refer to a dinosaur—

Behold now behemoth, which I made with thee; he eateth grass as an ox. Lo now, his strength is in his loins, and his force is in the navel of his belly. He moveth his tail like a cedar: the sinews of his stones are wrapped together. Behold, he drinketh up a river, and hasteth not: he trusteth that he can draw up Jordan into his mouth. (Job 40:15-17, 23)

If this scripture does refer to a dinosaur, such as a *Brontosaurus* or other sauropod, then there must have been at least one pair of young dinosaurs on Noah's ark!

The reader should also be informed that many instances of ancient art display images of dinosaurs.[48] Below is a sample list of such findings—

- A Cambodian monastery built approximately AD 1186 has a stone carving that is unmistakably a stegosaur.

- A Native American petroglyph, discovered in the Grand Canyon in 1879, depicts the image of what appears to be a trachodon.
- Aborigines of Australia etched a pictograph of a plesiosaur surrounded by numerous hunters. Its long neck and large flippers are clearly visible.
- Many figurines excavated from Mexican burial sites appear to be dinosaur-like creatures.
- Numerous burial stones excavated from graves in Peru have engravings that resemble dinosaurs, and some appear to be T. rex.
- A vivid Native American petroglyph of a brontosaurus is located under the Katrina Bridge at the Natural Bridges National Monument in the State of Utah.
- A petroglyph depicting a pterodactyl can be found at San Rafael Swell in Black Dragon Wash, Utah.

Evolutionists claim that these carvings were inspired by fossils discovered by their ancient sculptors. The author considers this very unlikely, however, since (*i*) all such fossils would need to be nearly complete skeletons, which are extremely rare, and (*ii*) the carvings resemble modern renditions of dinosaurs that were made using the advantages of modern technology. Furthermore, the nearest stegosaurus fossils to the Cambodian monastery carving are found in China—too far away to be seriously considered as a source of artistic inspiration.[49]

Another interesting dimension to the "age of dinosaurs" (from 201 million to 66 million years ago, according to evolutionists) is told in the fossil record of burmite, or Burmese amber. Burmite is found in present day Myanmar, formerly known as Burma. More than 3,000 burmite fossils have been found and photographed. A large deposit of these fossils has been collected by scientists such as Drs. George Poinar and Ted Pike. The fossils have been dated between 100 million and 110 million years, which positions them 34 million to 44 million years before the end of the age of dinosaurs. Burmese amber fossils include many "living fossils" (fossils of creatures still living today) that prove certain modern species actually lived during the age of dinosaurs.

These living fossils, encased in Burmese amber, include the following life forms—click beetle, weevil, moth, grasshopper, mayfly, caddisfly, lacewing, cockroach, bark beetle, walking stick, cicada, plant bug, bee, long horn beetle, ichneumonid wasp, gnat, midge, queen ant, praying mantis, centipede, millipede, jumping spider, tick, scorpion, garden spider, bark lice, leaf bits, nematodes (round worms), and a snail shell. *"The majority of these tiny entombed relics of the past look just like today's versions."* [50] The reader can see 167 specimens of beautiful and intriguing burmite fossils at the website—

http://home.fuse.net/paleopark/index.htm

Completeness of the Fossil Record

In order to assess how complete the fossil record was during his time (1985), Dr. Michael Denton—a British-Australian biochemist who is not a creation scientist but believes in intelligent design—used the listings in Dr. Alfred Romer's 1966 textbook *Vertebrate Paleontology* to calculate the percentage of living *orders* and *families* of terrestrial vertebrates that had been found in fossilized form as of 1985. Figure 25 (see the next page) shows the results of Dr. Denton's work.

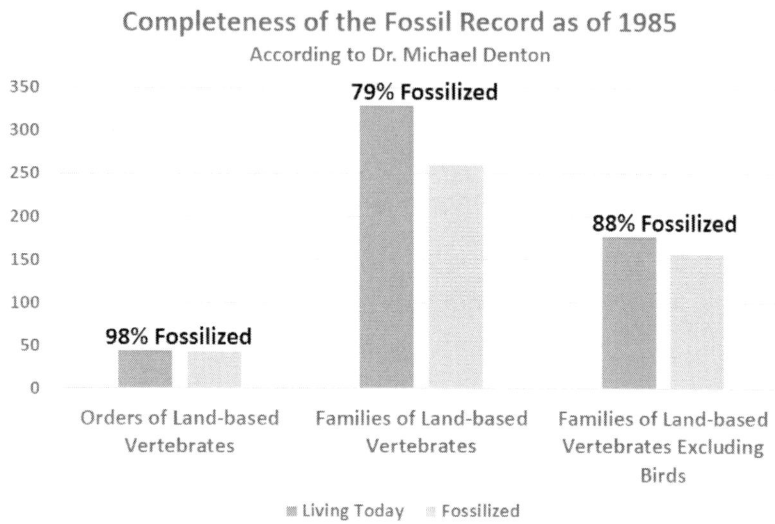

Figure 25—comparing Romer's 1966 List of Terrestrial Vertebrates with the Fossil Record of 1985

While the numbers from 1966 and 1985 shown in Figure 25 have undoubtedly changed since Dr. Denton made these calculations, the percentages of fossilized discoveries have probably gone up, since the number of uncovered fossil families has likely increased faster than the corresponding counts of known living terrestrial vertebrates (in fact, the known living counts may have not increased appreciably at all).

In any case, the work that Dr. Denton did shows that the fossil record is virtually complete with respect to basic types of land-based vertebrates.

Fossils of Special Interest

As discussed previously, when Charles Darwin published *On the Origin of Species by Means of Natural Selection* in 1859, he was concerned that the sudden appearance of numerous different animal groups in the fossil record, with no evidence of evolution from more primitive ancestors, could be raised as a serious objection to his theory of evolution. In the 1872 sixth edition of his book (where the title was first shortened to *The Origin of Species*), Darwin wrote[51] the following—

"Why, if species have descended from other species by insensibly fine gradations, do we not everywhere see innumerable transitional forms? Why is not all nature in confusion instead of the species being, as we see them, well defined? ... But, as by this theory innumerable transitional forms must have existed, why do we not find them embedded in countless numbers in the crust of the earth? ... Why then is not every geological formation and every stratum full of such intermediate links? Geology assuredly does not reveal any such finely graduated organic chain; and this, perhaps, is the most obvious and gravest objection which can be urged against my theory."

Starting with Darwin, evolutionists have been seeking fruitlessly for such transitional fossils. The initial false alarm came in Darwin's day, when the first fossil for *Archaeopteryx* was found in Bavaria. This fossil is still claimed by many to be the transitional link between reptiles and birds, even though other potential links have since been identified. None of the traits Archaeopteryx possessed are unique to either birds or reptiles. In reality, Archaeopteryx was a true bird—with a wishbone, perching feet, modern feathers, and complete wings. Although it had teeth, other extinct fossil birds did also, eliminating this trait as a potential transitional characteristic. The most likely ancestor of Archaeopteryx is dated 25 million years *younger* than Archaeopteryx. Knowledgeable evolutionists no longer consider Archaeopteryx to be the missing

Archaeopteryx fossil
courtesy Wikimedia Commons (in the public domain - expired copyright)

link between reptiles and birds, and yet it still is touted as a transitional form in many evolution training and marketing materials. For example, Wikipedia positions Archaeopteryx as—*"a genus of early bird that is transitional between feathered dinosaurs and modern birds"*—while also reporting—*"older potential avialans have since been identified, including* Anchiornis, Xiaotingia, *and* Aurornis.*"* [52]

While cladistics theory may utilize one or more of the highly questionable "linkage" fossils to create hypothetical bird family trees of the past, even greater importance is placed on theoretical creatures that have never been found. Once again, *"Evolution drives the bus, while the facts take a back seat."* [53]

A similar "just-so" story applies to the evolution of horses, where an impressive four-step transition path is frequently shown that begins with an extinct fossilized animal known as *Hyracotherium* (meaning "hyrax-like beast"). In the evolutionary path, the number of toes per leg goes from 4/3 (4 in the front, 3 in the rear for the hyrax) to 3/3 to 1/1 (in modern horses) while simultaneously "browsers" become "grazers" (with a different set of teeth). This upgrade path supposedly took 65 million years to occur.

Modern hyrax
courtesy Josski, Wikimedia Commons (in the public domain)

However, modern research has exposed some glaring challenges to this evolutionary story. First, the modern hyrax still lives today, proving that stasis occurred, not evolution. Second, fossils of the different transitional forms have been found in the same sedimentary rock strata—indicating that they lived at the same time and could not have had an ancestor/descendant relationship. Third, the amount of variation found within and between horse species that have adapted to their different environments through adaptive evolution is greater than the variation found in the fossil record that mainstream scientists call forward Darwinian evolution. Furthermore[54]—

"... the many breeds of horse produced by human intervention show wider variation than the natural species, but variation is not evolution."

Undoubtedly, the "missing link" of greatest interest is that for humans—the link between man and ape. Overzealous efforts to find this missing link have resulted in some of science's most shameful events. For example, consider the case of "Piltdown Man." Parts of a skull and jawbone were found near the English village of Piltdown in 1912. These fossils were promptly acclaimed as the evolutionary link halfway between ape and man, named *Eoanthropus dawsoni* (Dawn Man of Dawson, its discoverer). For more than forty years, this transitional link was considered by leading evolutionists all around the world to be the true evolutionary ancestor of modern man. Piltdown Man *"... was also used for decades to indoctrinate students with the idea that this fantastic find falsified the biblical view of mankind's special creation."* [55]

During the 1950s, a mineral mismatch discovered between one Piltdown Man bone and the rock from which it had been taken pushed scientists to take a second look at the revered fossils. Upon further analysis, the skull turned out to be a human skull with key parts missing. The jawbone was then determined to be from an orangutan whose teeth had been filed to look more human-like. Both had been stained to make them look more ancient and they had been buried with some other fossils, including one from Africa.

Piltdown Man was finally recognized for what it really was—a deliberate, shameful fraud!

"The evolutionary worldview had so strongly influenced the scientific research that for forty years observers had seen what was not there and failed to see what was there." [55]

Unfortunately, the impact of Piltdown Man was devastating before its fraudulent history was revealed—

*"While Piltdown Man was being touted as a 'fossil link' from ape to man in the early 1900s, Darwin's second book (*The Descent of Man, *1871) and the work of 'Germany's Darwin' Haeckel suggested to some evolutionists that Australian aborigines and African Negroes were subhuman 'living links.' Haeckel, who was caught mis-drawing human embryos to make them look more animal-like, drew aboriginals as animals in trees and claimed that they could not count as high as dogs or horses. Hitler based his beliefs on the racial inferiority of Jews and the superiority of Aryans* [Europeans and Western Asians] *on his interpretation of Darwinian ideas."* [55]

Piltdown Man *"... was also used for decades to indoctrinate students with the idea that this fantastic find falsified the biblical view of mankind's special creation."* [55]

Piltdown Man was not the only missing link fraud, hoax, or error that contributed, during the first half of the 20th Century, to widespread and misdirected faith in mainstream science's insistence that man evolved from ape. Some others[65] include—

- ***Nebraska Man***—a single fossil tooth discovered at Snake Creek, Nebraska in 1922 was given to Dr. Henry F. Osborn, head paleontologist at the American Museum of Natural History in New York. Osborn called for a complete reconstruction of the ape-man, based only on this tooth and his own aspiration for evidence of human evolution. Before long, illustrations of an ape-like subhuman began appearing in newspapers. However, 1927 excavations at the site exposed the truth that the tooth came from what was then considered an extinct peccary! Adding to the shame of this error, the infamous "Nebraska Pig" was discovered living in Paraguay in 1972!

- ***Java Man***—in 1891, a skull cap and three teeth were discovered on the Indonesian island Java by Dr. Eugene Dubois, who admired Ernest Haegel—the German biologist who marched lockstep with Charles Darwin. In 1892, Dubois found a thighbone (femur) 46 feet (14 m) from the location where the skull cap was found. Although the femur was clearly human-like, Dubois claimed until his dying day that he had found "the real missing link." However, in 1920 the truth emerged that Dubois had also uncovered two patently human skulls not far from, and at the same level as, Java Man's remnants. Dubois had kept this truth secret for nearly 30 years so he could promote Java Man. Discovery of the human skulls proved that Java Man was contemporary with humans, and thus could not be an evolutionary ancestor. *"While the shape of the skull would be unusual today, it falls within the range of human variability."* [65]
- ***Peking Man***—uncovered in China about 25 miles (40 km) from Peking during the 1920s and 1930s, much of the truth surrounding Peking Man was kept from the public while evolutionists claimed it was a missing link. Remains of ten obviously human natives were also uncovered at the site; thus Peking Man was also contemporary with humans, and therefore could not be a missing link ancestor. The original bones were mysteriously "lost" during the early 1940s. However, casts of them indicate that *"Peking Man was a bit smaller overall than the average human today, but in all respects falls within the modern range of* [human] *variation."* [65]

In 1925, the famous Scopes Monkey Trial was conducted in Dayton, Tennessee. High school teacher John Scopes was on trial for teaching evolution, thus violating Tennessee's Butler Act which prohibited the teaching of evolution in any state-funded school. Scopes was found guilty and fined $100, but the verdict was overturned on a technicality. The ACLU soon after took on the case as a worthy cause, but only until 1932.

How things have changed since 1925! Now the ACLU is biased against Christianity and the courts have ruled that only evolution—not creation—can be taught in government-funded schools!

Despite media publicization of the Piltdown Man fraud in 1953, the 1958 United States National Defense Education Act brought about new biology textbooks, produced in cooperation with the American Institute of Biological Sciences, that stressed the importance of evolution as the unifying principle of biology. By 1967, the Butler Law had been repealed in Tennessee. The reader will undoubtedly recognize how far the pendulum has swung to the side of evolution since 1925!

Following the Piltdown Man deception, several other proposed human ancestors have come and gone. Currently, the two most popular fossils that are claimed to link man to apes are Neanderthal and "Lucy."

Neanderthal bone fossils were initially collected during the mid-1800s. They were found in the Neander Valley, which is a limestone canyon located several miles east of Dusseldorf, Germany. The original Neanderthal discovery occurred during a mining operation in the canyon. The bones were found in a cave that no longer exists. At first, Neanderthal bone fossils were merely a curiosity. However, they took on more importance during the early 1900s, thanks to Piltdown Man. Neanderthal has been a subject of thorough study ever since.

Neanderthal skeletons reveal a somewhat stocky build and a sloping forehead, but otherwise they mirror modern human skeletons. Although their habitations clearly indicate a primitive lifestyle, Neanderthals lived in communities and raised crops, hunted large animals using weapons, buried their dead and observed religious rites, built musical instruments that use the modern scale, and at least thought about an afterlife. Thus Neanderthals were fully human, not transitional sub-humans. They lived primarily in the northern parts of Europe that were impacted by the post-Flood Ice Age. There they found caves to be the safest and warmest places to live. Although their lifestyle was meager, they were intelligent enough to survive in the harsh conditions caused by the Ice Age.[56] Perhaps they were a family group that had migrated from the Tower of Babel at the time the Lord confounded the language of the people.

In July 2006, the Neanderthal Genome Project was launched.[57] Coordinated by the Max Planck Institute for Evolutionary Anthropology in Germany and 454 Life Sciences in the United States of America, this focused collaboration of scientists was put in place to sequence the Neanderthal genome.

DNA Structure
courtesy Richard Wheeler

As a brief background, the DNA sequence of an organism encodes all the biological information required for that organism to survive and reproduce. DNA sequencing is the process of determining the exact order of nucleotides within a DNA fragment. Referring back to Chapter 2, nucleotides make up the individual steps of the spiral staircase structure of DNA, shown here. Each nucleotide is composed of a *base*, a sugar molecule, and at least one phosphate molecule. The four possible bases are A (adenine), C (cytosine), G (guanine), and T (thymine). *Base pairs* (G-C and A-T) form naturally according to hydrogen bonding patterns, allowing the DNA helix to maintain its regular helical structure independent of the sequence of its nucleotides. The complementary nature of this amazing base pair structure provides a backup copy of all the biological information encoded within the double-stranded DNA molecule.[58]

The initial draft of the Neanderthal genome study, based on four billion base pairs of Neanderthal DNA, was published in the May 2010 edition of the journal *Science*. This study found that mixing of genes had occurred between Neanderthals and modern, non-African humans.

The entire Neanderthal genome was then published in December 2013. The genome was extracted from 21 bone samples that came from Siberia, Croatia, Spain, and Germany. Preliminary sequencing results[57] showed that *"99.7% of the base pairs of the modern human and Neanderthal genomes are identical, compared to humans sharing around 98.8% of base pairs with the chimpanzee."*

Recently, Dr. Jeffrey Tomkins, a genetics research associate at the Institute for Creation Research, did his own comparison of Neanderthal and modern human DNA sequences. Using publicly available data sets obtained from the Max Planck Institute which contained millions of individual Neanderthal DNA sequences, he found that, **on average, 99.7% to 100% of the modern and ancient sequences were identical. These preliminary results, along with those from the Neanderthal Genome Project, clearly indicate that Neanderthals were human beings, not transitional sub-human ancestors.**[59]

Regarding the quote above, *"… compared to humans sharing around 98.8% of base pairs with the chimpanzee,"* other studies by evolution scientists lowered the commonality between man and chimpanzee from 98.8% to 94%—showing that the genetic gap between humans and chimps is far larger than originally thought.[57] However, Dr. Tomkins did his own research in 2013 to verify the 94% number. To his amazement, he found that levels of human-chimp DNA similarity were significantly lower than commonly reported (usually 98.8%). In fact, after comparing about 40,000 chimpanzee genomic sequences against the human genome, he found that only 69% of the chimpanzee X chromosome, and only 43% of its Y chromosome, were similar to human chromosomes.

Genome-wide, only 70% of the chimpanzee DNA was similar to human DNA under the most optimal alignment conditions. Evidently, evolution scientists had been cherry picking only those portions of the two genomes that are highly similar in order to support their claims of 98.8%, or even 94%, identical base pairs.

Dr. Tomkins' conclusion[60] was—

"While chimpanzees and humans share many localized protein-coding regions of high similarity, the overall extreme discontinuity between the two genomes defies evolutionary time-scales and dogmatic presuppositions about a common ancestor."

Bottom line, the Neanderthals were merely an ethnic group within the human race, and were even closer to the norm of all ethnic groups than are modern Australian Aborigines.[56]

The second fossil which evolutionists hold up as the first human, or the ape that took the first step to become human, is known as "Lucy." Her bones and bone fragments were discovered in Ethiopia by Dr. Donald Johanson in 1974. They were gathered from a rather large area, and added up to about 40% of a complete skeleton. Only the lower half of her jawbone, a single hip, the sacrum, one femur, and several ribs were intact. The complete extent of her remains is shown as a cast supported by a steel frame in Figure 26 on the next page. No feet were found, and only one small hand bone was discovered. Therefore, some of the most important clues to Lucy's frame and her way of life were missing. Her official Latin name is *Australopithecus afarensis*—meaning "southern ape from the Afar region."

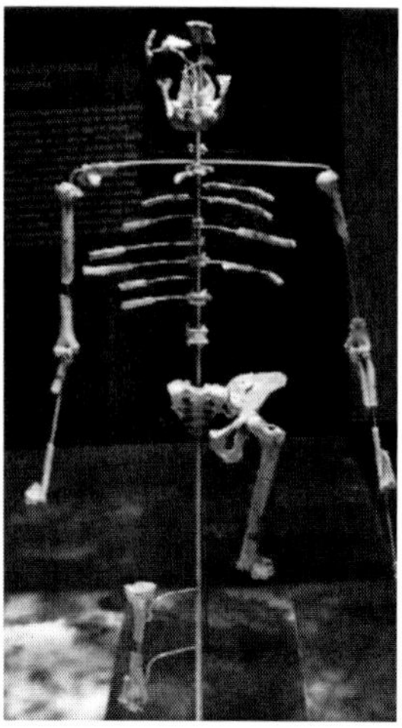

Figure 26—Cast of Lucy in Mexico City
courtesy Danrah, Wikimedia Commons (in the public domain)

Since Lucy's discovery, several more bones, which are believed to be from the same species, have been discovered. By putting them all together, a more complete representation of Lucy has emerged. She stood only 3½ feet (1 m) tall (typical of modern chimpanzees) and was very chimp-like—especially from the waist up. Her jaw was clearly V-shaped, not U-shaped like a human jaw, and her hyoid bone, located in the neck to aid vocalization, was also chimp-like. Furthermore, Lucy's teeth were chimp-like. Perhaps more important, her shoulder blade resembled that of a young chimpanzee or gorilla—indicating that she partially supported her weight by walking on her knuckles and keeping her spine nearly horizontal. Lucy's shoulder blade also positioned her arms so they could be easily lifted overhead for tree climbing.

Similar fossils of other Australopithecines were found in the same region. They have long, curved fingers and toes, with an opposite "thumb" on the feet to aid swinging from tree branches. Dated more recent than Lucy, one would expect these fossils to be more evolved, but they are clearly ape-like.

Not far from the region where Lucy's bones were found is a place called Laetoli. Both locations are dated approximately the same (3.2 to 3.6 million years). In 1978, archeologist Mary Leakey excavated the Laetoli location, uncovering human-like footprints that were preserved in volcanic ash. Because these fossilized footprints are identical to modern human footprints, they could not have been made by a creature that has an opposable big toe. However, evolutionists insist that they were made by Lucy's species—**even though they could not have been made with a**

chimp-like foot that was common to the australopithecine group. Most likely, the Laetoli prints were made by humans.

According to paleontologists, an upright stance is as a critical identifying feature of humans. Only two parts of Lucy could possibly support Dr. Johanson's claim that Lucy walked with an upright stance—her knee and hip. Remarkably, Johanson found the knee a year before he discovered Lucy, and it was located in an entirely different region. However, he felt that anatomical similarities justified including it with the other bones that comprise Lucy. However, others have disagreed. Regardless, and more importantly, professional anatomists have disagreed with Johanson's belief that Lucy's knee allowed her to walk uprightly like a human.

In mammals, the hip bone determines how legs are attached and oriented, and also affects a creature's stance and gait. When Dr. Johanson came to the conclusion that Lucy walked upright, the hip he had examined was only one side of Lucy's pelvis. To date, no other Australopithecus pelvis has been found. However, the nearly complete body of an Australopithecus infant was found recently, and it is *"grossly ape-like."*[61] Since infant chimpanzees are known to undergo significant body changes as they grow, one cannot use the infant pelvis of this more recently discovered fossil to conclude that Lucy could or could not walk upright.

Over the years since finding Lucy, Dr. Johanson and his colleagues have made numerous additional claims. In 1994, a documentary of Dr. Johanson's discoveries was made by the Public Broadcasting System (PBS). It became a three part NOVA series titled *In Search of Human Origins*. In episode one ("The Story of Lucy"), Johanson admitted that Lucy's partial pelvis, as initially found, resembled a chimpanzee pelvis, not capable of supporting erect walking. At that time, the crushed half pelvis consisted of forty different pieces. Johanson then added that he would have to reconstruct the pelvis in order to determine Lucy's posture. His colleague, Dr. Owen Lovejoy, "came to the rescue." Lovejoy first separated all the pieces so he could reassemble them in proper order—like a three-dimensional jigsaw puzzle. However, when he tried to do this, the pieces did not fit together to form the configuration he wanted. Rather, they readily came together in the configuration of a chimpanzee's pelvis.

What happened next in this PBS documentary is reproduced below from the actual transcript that was formerly posted on the PBS website—

> DON JOHANSON: We needed Owen Lovejoy's expertise again, because the evidence wasn't quite adding up. The knee looked human, but the shape of her hip didn't. Superficially, her hip resembled a chimpanzee's, which meant that Lucy couldn't possibly have walked like a modern human. But Lovejoy noticed something odd about the way the bones had been fossilized.
>
> OWEN LOVEJOY: When I put the two parts of the pelvis together that we had, this part of the pelvis had pressed so hard and so completely into this one, that it caused it to be broken into a series of individual pieces, which were then fused together in later fossilization.
>
> DON JOHANSON: After Lucy died, some of her bones lying in the mud must have been crushed or broken, perhaps by animals browsing at the lake shore.
>
> OWEN LOVEJOY: This has caused the two bones in fact to fit together so well that they're in an anatomically impossible position.
>
> DON JOHANSON: The perfect fit was an illusion that made Lucy's hip bones seem to flair out like a chimp's.

> But all was not lost. Lovejoy decided he could restore the pelvis to its natural shape. He didn't want to tamper with the original, so he made a copy in plaster. He cut the damaged pieces out and put them back together the way they were before Lucy died. It was a tricky job, but after taking the kink out of the pelvis, it all fit together perfectly, like a three-dimensional jigsaw puzzle. As a result, the angle of the hip looks nothing like a chimp's, but a lot like ours. Anatomically at least, Lucy could stand like a human."

"In the film, Lovejoy is seen using a grinding wheel to cut the pieces, shape them, and place them so that they mesh into a human-like form rather than the original ape-like form. He justifies this by using the analogy of a broken teacup, which can be glued back together by one who knows how. But this analogy presupposes that one knows that the fragments were, in fact, from a broken teacup. It presupposes that he knows the correct original shape of the teacup, that none of the pieces are missing, and that all the fragments being assembled are actually from a single teacup. His reconstruction of the shattered hipbone—the type of which had not been found elsewhere, and thus had an unknown original shape—has been the basis of years of evolutionary speculation and has spawned inferences on Lucy being 'the first human' or the first step toward humanity. But what if the reconstruction, which was so intricate and extensive, resulted in unintended error?" [62]

The author encourages the reader to decide for himself/herself whether the claims made about Lucy walking like a human are founded on scientific principles or not.

Chapter Summary and Conclusions

This chapter has provided scientific evidence, or other basis, for the following truths about the fossil record—

- The worldwide Cambrian Explosion fossil record was created by the catastrophic Global Flood.

- Essentially no trace of transitional life forms, required for forward evolution, can be found anywhere in the fossil record. Thus, Darwin correctly pointed out a fundamental fallacy of his own theory.
- When fossils first appear in the rock strata, they are fully formed and functional.
- The evolutionary concept of punctuated equilibria has no scientific basis. Geneticists are justified in rejecting it as an evolutionary hypothesis.
- The reality of stasis (little or no evolutionary change) is manifested throughout the fossil record.
- Mainstream science no longer looks to the fossil record for evidence to support its latest theories of evolution.
- Cladograms are graphical illustrations of an individual researcher's hypotheses. They are not fact, and do not provide scientific proof of forward evolution.
- Distribution of the fossil record (95% marine invertebrates, < 5% plants, < 1% vertebrates) mirrors the distribution predicted by the Global Flood.
- The global fossil record—recorded in sedimentary strata (some of which cover hundreds of thousands of square miles)—could only have been deposited by the Global Flood.
- The typical mixture of fossil types contained within individual strata could only have been deposited by the Global Flood. The typical mixture includes terrestrial life forms, marine life forms, and vegetation from various environments.
- Fossils of animals preserved in the act of swallowing other animals, or giving live birth, provide reliable evidence that these animals were rapidly buried in a liquid grave.

- Numerous fossils of animals preserved in the act of gasping for breath provide solid evidence that these animals drowned and were rapidly buried in a liquid grave.
- The fact that scientists have discovered dinosaur mummies and other fossils harboring soft tissues, squid ink, and even blood cells cannot be denied.
- No reliable scientific explanation has been given for the existence of such mummies, soft tissues, squid ink, and blood cells in the fossil record. Logic dictates that this evidence proves dinosaurs are not millions of years old; however, modern science will never adopt this explanation since it does not support current theories of evolution.
- Many examples of ancient art displaying images of dinosaurs have been found around the world. Christians understand that the ancient artists who created these works of art lived during the past 6,000 years—after the Fall of Adam and Eve.
- Preliminary results produced by the Neanderthal Genome Project show that the Neanderthal genome essentially matches that of modern humans.
- The theory that "Lucy" was able to walk upright like a human is not based on reliable scientific evidence; rather it is based on a fossilized pelvis that was deliberately modified and then pieced together so as to look like a human pelvis.
- Many creatures have been found living today that were thought to have been extinct for millions of years, and their bodies closely resemble those of their fossilized ancestors. Darwin called these "living fossils."
- The fossil record is essentially complete.

Charles Darwin was among the first to realize that the fossil record does not support his theory of evolution by mutation and natural selection. Over the years, the expanding fossil record has not been kind to Darwin—it still shows that forward evolution did not happen in the past. And now the fossil record is essentially complete.

Unfortunately, modern science has downplayed and even ignored the fossil record in order to sustain its allegiance to uniformitarianism and evolution. This chapter has shown how mainstream science has had to continuously develop new theories and hypotheses in order to deal with fossil record facts that conflict with its tenets. Furthermore, any scientist who does not conform to modern science doctrine is ostracized, and if a scientist dares to bring the revealed word of God into his or her work, that work is branded *pseudoscience.*

Modern science is quick to promote examples of evolution at work today that are easily documented and explained in order to remain relevant in modern society. However, the examples they highlight are invariably those of adaptive evolution—not forward evolution. An uneducated public is thus easily convinced that evolution, as universally taught by the media, the Internet, public education, mainstream contemporary science, and governments, is truth.

The authors of *The Fossil Record* summarized the tension between mainstream science and the breakthrough science presented in this chapter as follows—

"And there is the problem—when you step back and take a look, the data just do not add up for the 'mainstream' view of fossils. Evolutionary 'just so' stories may be entertaining and imaginative, but they consist of inferior historical reconstructions of the unobserved past. Stories should not be allowed to substitute for evidence. Evolution is neither the only

possible story, nor the best possible story. It is simply not a credible explanation of the evidence left by the unobserved events of the past."[63]

"With or without its favorite stories, evolution remains the dominant origins myth for Western culture, modifying itself to fit all observations. If the theory can accommodate any possibility, it is a weak concept indeed, full of sound and fury that signify nothing."[64]

Christians with an open mind should, after reading this chapter, realize that the fossil record supports neither Darwinian evolution nor the geologic timescale. Rather, it does support the truths that have been revealed to the prophets—specifically that God created all living things, that they reproduce after their own kind, and that they, like man, have existed on this mortal earth for a period of about 6,000 years.

Had the Global Flood never occurred, there would be no sedimentary strata to form earth's geologic record for the Paleozoic, Mesozoic, and Cenozoic eras. Nor would there be a fossil record within these strata. When living organisms die, their remains rapidly disappear through consumption by scavengers, decay, decomposition, and disintegration. Thus uniformitarianism would have resulted in an earth that has essentially no geologic or fossil records above the Great Unconformity. Only the catastrophic, global Flood could have created the geologic and the fossil records that exist today. These records truly testify that the Global Flood cleansed the earth of its former violence and wickedness.

As it turns out, Leonardo da Vinci provided mankind with breakthrough science in the 1500s when he proposed that fossils are the remains of animals that were buried in soft sand and mud! How tragic that modern scientists refuse to accept it.

The author of this book is not the only author who has run into pseudoscience originating with mainstream science. In the book *Lost American Antiquities: A Hidden History*, author S. Edgar Smoot wrote [66] the following—

"... there is a politically correct line that many scholars knew of, which they were afraid to step over. ... What we learned by asking such questions was that there were mainstream historians, anthropologists, and archaeologists who did not want to upset the proverbial apple cart of academic tradition. They understood well that they had to tailor their comments, in an effort to be politically correct. ... As a result, there has been a vast amount of pseudoscience that has been handed down to our day. Not only by those outside the employment of the sciences, but from those who have been hired from within. These guardians of the gate into the fields of the evolutionary sciences understand well, the implications of how a view into the origin and evolution of man can change how society views and values God, faith based organizations, children, marriage, family and even the founding documents of America."

Let the reader beware of accepting mainstream science at face value!

CHAPTER FIVE

Putting It All Together

A great deal of information has been presented in this book—perhaps even an overwhelming amount. The author's intent for sharing all this information has been to convey the truth that

forward (Darwinian) evolution does not happen—ever!

A high level recap of the book may help the reader remember and organize the key points that were made in order to arrive at this vital truth—***which increases faith in God as The Creator!***

Chapter 1 Recap

This chapter discussed some laws and conditions of mortal life on earth—

1. All living things will eventually die; therefore, they have a beginning and an end.
2. All living things can only bring forth after their own kind.
3. Everything in mortality naturally degenerates; hence the cause of death.

Modern science has formally restated number 3 above as the Second Law of Thermodynamics. This law is about entropy, which is best understood with the analogy of a wind-up clock—

nothing happens until the clock is wound; it then proceeds to keep time as the spring mechanism winds down. When the mechanism is totally unwound, the clock stops functioning.

Chapter 1 also revealed that a significant area of disagreement between scripture and mainstream contemporary science is the extent of the Flood. Most scientists believe the Bible's account of the Flood is a myth.

Chapter 1 then reviewed some laws and conditions of eternity—

- There are different laws and conditions that apply to the heavens and the mortal earth.
- Jesus Christ abolished death through the resurrection—which replaces mortality with immortality. All who have lived on this earth, or who will yet live on earth, will be resurrected with an immortal body.
- Christ's faithful followers will dwell with Him in His Father's house after they are resurrected. This promise implies that mortality will no longer apply after the resurrection; immortality will reign instead—confirming that there are truly different laws and conditions which govern life in different parts of the universe. How wonderful and glorious to contemplate that the laws and conditions of mortality are merely a temporary step to eternal life!
- No mainstream scientist has any inkling about the way God's universe actually functions. The only paradigm for the visible universe such a scientist can imagine is limited by the laws and conditions of mortality that determine his/her own existence.
- The fact that different laws and conditions apply to mortality and immortality is unknown to mainstream contemporary science; therefore, Christians should keep in mind that the

theories and hypotheses of modern science do not reflect the realities of God's universe that are revealed in the Bible.

Chapter 2 Recap

This chapter introduced two kinds of evolution—adaptive and forward.

Adaptive evolution utilizes the adaptive genetic capabilities that were given to all living things when they were created. Such adaptive capabilities allow organisms to evolve within a *kind* in a way that allows them to better survive in a given habitat. Adaptive evolution is God's way of ensuring survival of his living creations in a dynamic and hostile mortal world. Adaptive evolution was especially important after the animals on Noah's ark left their wooden, seafaring home and began to repopulate the earth.

Darwin believed that natural selection can bring about forward evolution. Forward evolution supposedly results in the gradual appearance on earth of new species that are more complex, or a higher order, than the species from which they evolved. While adaptive evolution only brings about changes within a kind (genus or family) of organisms, forward evolution is believed by mainstream science to bring about entirely new species that are genetically superior to the species from which they evolved, and hence are new kinds that did not exist previously. The genetic superiority supposedly results from the creation of new and additional genetic (biological) information (DNA).

Through forward evolution, man is thought to have ultimately descended from the most primitive form of life, which somehow gained its life from self-replicating but nonliving molecules.

Forward evolution requires entropy to decrease (or order to increase), breaking the Second Law of Thermodynamics. The theory of forward evolution involves random genetic mutations and extremely long periods of time for natural selection to do its work. Forward evolution is positioned by its champions to be the result of a pointless, unguided, amoral, natural process. There is no room in this characterization for God to guide forward evolution; indeed, there is no room in this characterization for God to even exist.

The fundamental basis upon which forward evolution is built is labeled by some scientists "The Primary Axiom." An axiom is defined as an established premise, principle, or rule, or a self-evident truth. *The Primary Axiom is that man is merely the product of random mutations and natural selection.*

Christians should understand that mainstream contemporary science is religiously—even fanatically—committed to forward evolution.

The road that forward evolution has taken since Darwin has not been smooth; rather, it has required "absurd constructs" and "unsubstantiated just-so stories" to keep it moving over the bumps that have been encountered as new science has emerged over the years. The bumps provide a solid foundation for proving that the Primary Axiom is false. They include—

- ✓ Haldane's Dilemma
- ✓ Kimura's Quandary
- ✓ Muller's Fear
- ✓ Muller's Ratchet
- ✓ Neel's Realization
- ✓ Kondrashov's Question
- ✓ Kondrashov's Numbers
- ✓ Crow's Concerns
- ✓ Lynch's Mutation Meltdown

- ✓ Higgins and Lynch – More Mutation Meltdown
- ✓ Hoyle's Big Picture

A world famous geneticist, Dr. John C. Sanford, became an atheist in his youth after studying Darwinian evolution. However, he began to question the Primary Axiom late in his career. In Dr. Sanford's own words—

"I did this with great fear and trepidation … To question the Primary Axiom required me to re-examine virtually everything I thought I knew about genetics. This was the most difficult intellectual endeavor of my life … What I eventually experienced was a complete over-throw of my previous understandings. Several years of personal struggle resulted in a new understanding, and a very strong conviction that the Primary Axiom was most definitely wrong. More importantly, I became convinced that the Axiom could be ***shown*** *to be wrong to any reasonable and open-minded individual. This realization was exhilarating, but again—frightening. I realized that I had a moral obligation to openly challenge this most sacred of cows. In doing this, I realized I would earn for myself the intense disdain of most of my colleagues within academia—not to mention very intense opposition and anger from other high places.*

"What should I do? It has become my conviction that the Primary Axiom is insidious on the highest level—having catastrophic impact on countless human lives. Furthermore, every form of objective analysis I have performed has convinced me that the Axiom is clearly false. So now, regardless of the consequences, I have to say it out loud: ***the Emperor has no clothes!****"*

True to his moral obligation, Dr. Sanford wrote a book titled *Genetic Entropy & The Mystery of the Genome.* This books contains breakthrough science that exposes the truth that the Primary Axiom is wrong.

Chapter 2 covered in some detail the scientific evidence provided by Dr. Sanford to prove that the Primary Axiom is false. In Dr. Sanford's own words—

"We have reviewed compelling evidence that even when ignoring deleterious mutations, mutation/selection cannot create a single gene—not within the human evolutionary timescale. When deleterious mutations are factored back in, we see that mutation/selection cannot create a single gene—ever. This is overwhelming evidence against the Primary Axiom. ***In my opinion, this constitutes what is essentially a formal proof that the Primary Axiom is false.***

Chapter 2 also pointed out that Dr. Sanford is not the only scientist who believes that Darwinian evolution is a false theory. As of June 2015, 946 scientists holding doctorate degrees (either Ph.D. or M.D.) had signed *A Scientific Dissent From Darwinism.*

Dr. Sanford also wrote the following about adaptive evolution—

"In conclusion, the genome must have been designed, and could not have evolved. Yet we know that 'micro-evolution' (adaptive selection) does happen, correct? How can this be? ... I believe most useful variation must be ***designed****. When we see adaptive selection occurring, we are usually witnessing segregation and recombination of useful variants of genes and gene components—***which were designed to segregate and recombine in the first place****. We are not usually seeing the result of random mutations—which are consistently deleterious. Selection operates to eliminate the worst of mutations, while favoring the most desirable recombinants and segregants of designed variation. ... It is this type of designed diversity that natural selection can act upon most effectively. All such designed variants would be expected to be created within useful linkage groups, and would have originated at high allelic frequencies.*

... Functional linkage groups and high allele frequency allow for very rapid responsiveness to selection, and thus rapid local adaptation."

Dr. Sanford's work shows that, instead of continually advancing through forward evolution, the genomes of all living things are actually deteriorating. A genome is the complete set of an organism's hereditary information—its book of life, the sum of an organism's genetic parts, including all of its genes and chromosomes. Dr. Sanford called this deterioration genetic entropy. Thus he showed that the genomes of living things obey the Second Law of Thermodynamics, just like everything else in mortality.

Remarkably, mainstream genetic science acknowledges the fact that the human genome is deteriorating, but does not accept the logical conclusion that the Primary Axiom is false.

For the author of this book, the strongest evidence of the reality of genetic entropy is gleaned by plotting the lifetimes of early mankind, as shown in Figure 8. Dr. Sanford was evidently raised up by God to provide scientific proof that forward evolution cannot happen.

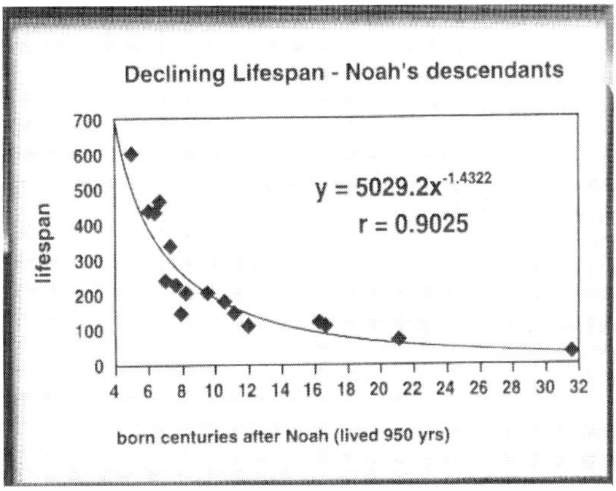

Figure 8

Chapter 3 Recap

This chapter introduced the geologic time scale (see Figure 9 on the next page) that is embraced by mainstream contemporary science; also by the United States Geological Survey, and by the International Commission on Stratigraphy. This time scale is organized to correspond with changes in the composition of the earth's strata (i.e., layers of rock with internally unique but consistent characteristics). The geologic time scale sets the age of the earth at 4.55 billion years (4550 million years), and shows that life began about 4 billion years ago.

Dinosaur fossils are found in the strata corresponding to the Mesozoic era. Radioactive dating methods are used to determine the age of rocks, fossils, archeological artifacts, and the earth itself to create this geologic time scale. A two million year Quaternary period (not shown), which is the time that recognizable humans have been on the earth, is too small to be visible at the scale of this clock.

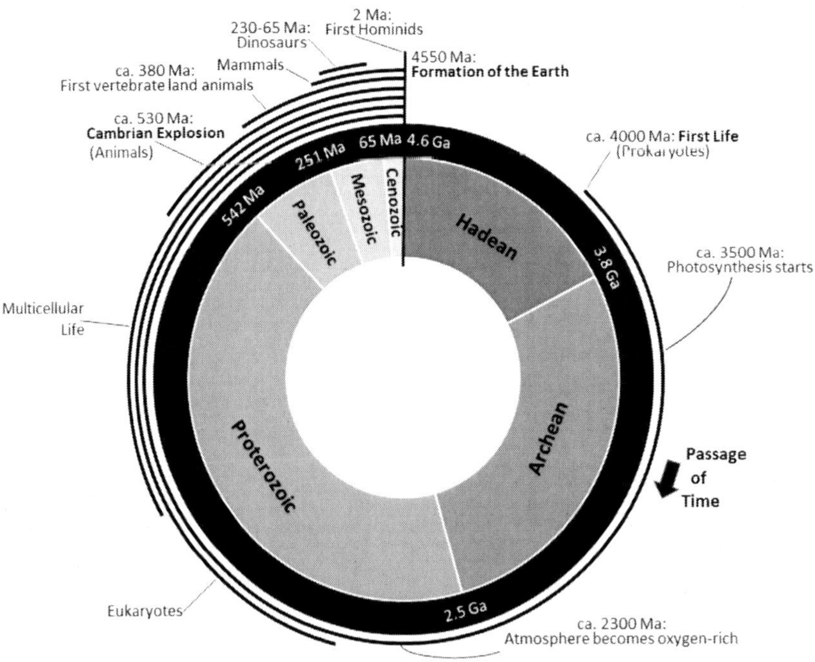

Figure 9—Geologic Time Scale Shown as a Clock
(Ma = 1 million years, Ga = 1 billion years)

Chapter 3 also discussed breakthrough science challenges to the geologic time scale. The following points were covered—

1. As with Darwinian evolution, mainstream science is committed to uniformitarianism and the geologic time scale shown in Figure 9.
2. The measured rate of helium diffusion in zircon crystals from deep underground granite was compatible with the crystals being about 5,000 years old—yet uranium-lead dating of the same granite resulted in an age of 1.5 billion years.
3. There are more than a dozen known locations where historical volcano lava flows were dated millions of years older than their actual ages, proving that dating methods involving argon are not reliable.
4. Using the high precision accelerator mass spectrometer (AMS) carbon dating method during the last twenty seven years, no fossils have been found anywhere in the world that are dated as old as 90,000 years. Rather, most fossils seem to have approximately the same $^{14}C/C$ ratio, which indicates an age of roughly 50,000 years. However, 50,000 years is too old because the rate of ^{14}C production in the atmosphere actually exceeds the rate of ^{14}C decay by 30%. Using other radioactive dating methods, these same fossils are dated up to 542 million years.
5. AMS carbon dating results imply that all fossilized organisms lived on the earth at the same time.
6. The recent discovery of polonium radiohalos deep inside samples of granite indicate the granite must have solidified in a very short time—only days, not millions or billions of years.
7. Recent studies of the earth's decaying magnetic field indicate that the mortal earth cannot be more than 9,000 years old.
8. Erosion rates of the earth's land surfaces do not support the hypothesis that the mortal earth is hundreds of millions or billions of years old.

Taken together, perhaps the most important truth to emerge from these points is that science cannot reliably date the age of the earth, nor the geologic time scale, nor the fossil record.

New science involving the interior of the earth indicates that water is found *everywhere* inside the earth. Even if only one percent of the earth's core is water, that will amount to ten times more water than is presently found in all the oceans. Scientists have found evidence of a vast water reservoir beneath eastern Asia that is at least the volume of the Arctic Ocean. The earth's transition zone might have as much water as all the world's oceans put together. One reason the earth is such a dynamic planet is the presence of water in its interior.

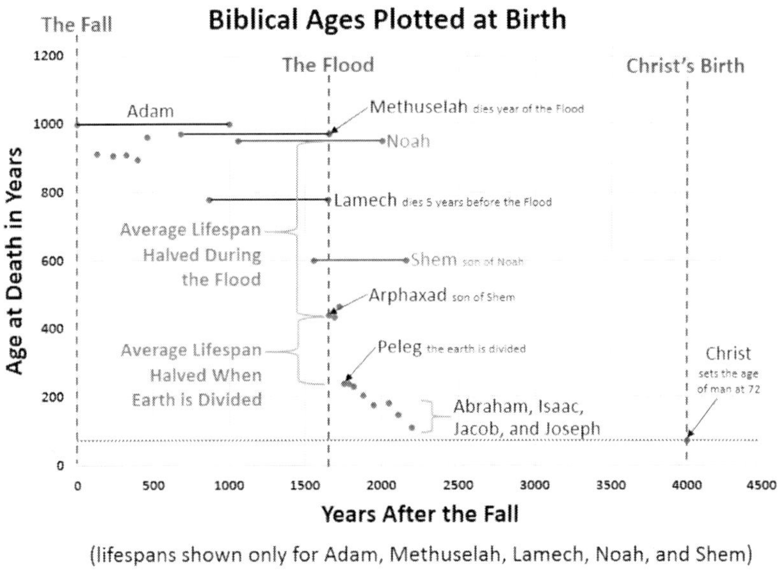

Figure 11—Biblical ages showing declining lifespans with two different catastrophic reductions caused by the Flood and the Division of the earth. Average lifespan was reduced by half as a result of these catastrophic events.

Chapter 3 also emphasized that the earth has had a catastrophic past that cut the average lifespan of mankind in half two different times, as illustrated in Figure 11 on the previous page.

The Flood, although it only lasted for about a year, cut the average lifespan of mankind in half, and this happened 1657 years after the Fall. Likewise, the Division of the earth that occurred in the days of Peleg cut the average lifespan in half once again—approximately a century after the Flood. Clearly, something catastrophic must have been going on with the earth during these two historical events in order to cause such drastic reductions in the average lifespan of mankind.

Chapter 3 also introduced the concept of catastrophic plate tectonics as the natural mechanism that drove the Global Flood. A breakthrough paper titled *Catastrophic Plate Tectonics: A Global Flood Model of Earth History* pulled together the work of many scientists to develop a state-of-the-art model based on rapid plate tectonics that explains geological evidence which had not been satisfactorily explained by conventional plate tectonics theory, nor by any other earth history model.

Virtually all of the pre-Flood ocean's lithosphere (the earth's crust and its upper-most solid mantle) was subducted during the Flood, essentially being recycled back into the earth's mantle. Present age-dating of the earth's ocean floors confirms this catastrophic phenomenon. The universal subduction was caused by a thermal runaway instability that enabled yards-per-second subduction of the ocean's lithosphere. This cataclysmic runaway was responsible for breaking up all the fountains of the great deep and for opening the windows of heaven.

A breakthrough computer simulation of the thermal runaway instability was achieved by Dr. John R. Baumgardner in 2003. Figure 17 on the next page shows a snapshot of the 2D simulation which represents the runaway 20 days after subduction began. Note in the top panel that runaway magma plumes emerge from both top and bottom boundaries. The release of gravitational potential energy stored in both of these boundary layers drives the circular motion.

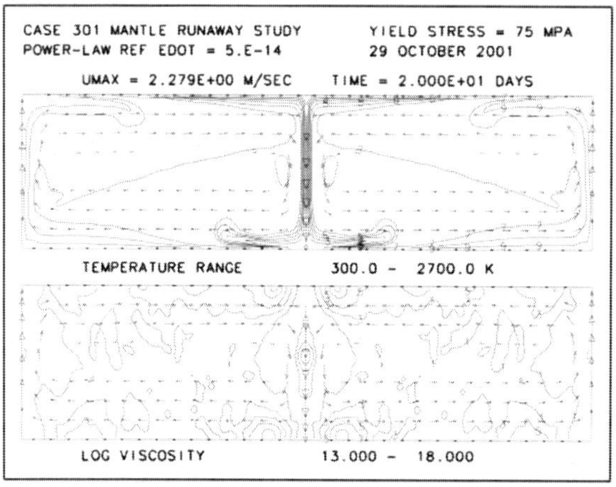

Figure 17—snapshot of thermal runaway instability within the earth's mantle 20 days after it began
courtesy Dr. John R. Baumgardner

Dr. Baumgardner used the results of this 2D computer simulation to take the next step and simulate a 3D view of what the runaway subduction event did to the pre-Flood supercontinent. Two snapshots of this 3D computer simulation are included on the next page. Figure 18 shows the present-day continents beginning to separate, and Figure 19 provides a snapshot of the moving continents 10 days later.

Seeing the results of catastrophic plate tectonics in this manner provides a clearer understanding of the Flood and its devastating effects.

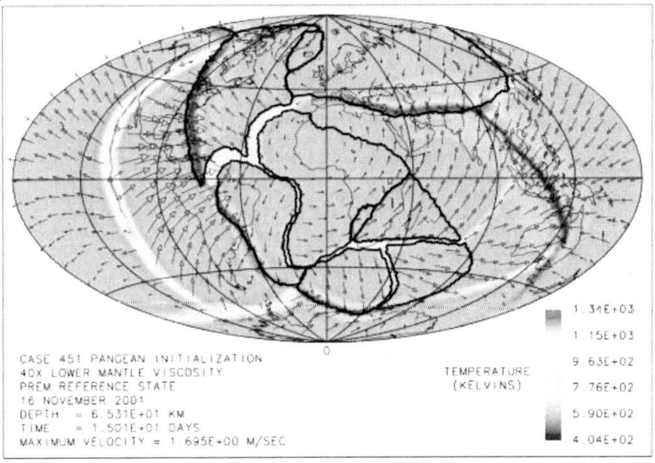

Figure 18—snapshot of the fracturing Pangaea supercontinent midway through the Flood (taken 15 days after the 3D simulation started)
courtesy Dr. John R. Baumgardner

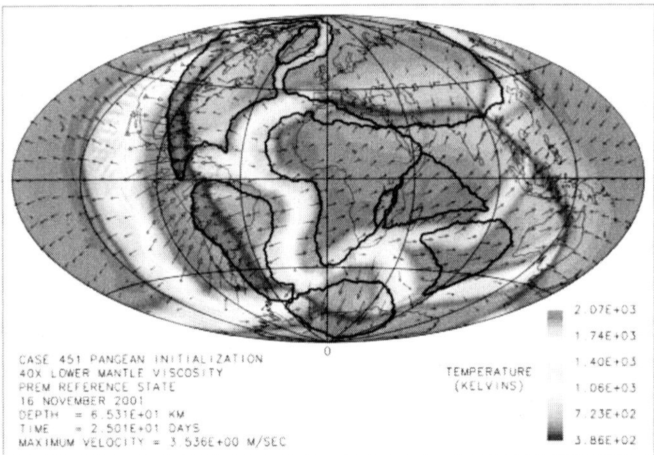

Figure 19—snapshot of 3D simulation results 10 days later, illustrating how the continents were pulled apart during the Flood
courtesy Dr. John R. Baumgardner

The first 150 days of the Flood caused the eventual fossilization of trillions of sea-dwelling organisms and the destruction of all land-based and flying creatures. First to be impacted were those organisms dwelling on the bottom of the ocean, especially near the erupting fountains. Then organisms within the shallow continental shelf areas were overcome as the initial deadly waves and underwater surges inflicted their tremendous forces. When the huge waves broke over the coastline, anything dwelling there was smothered and eventually buried in the watery sediments that were transported by the raging sea. As the pulsing walls of water climbed ever higher and farther over the supercontinent, all life within the low-lying areas, and finally the upland regions, died a horrific death. The Flood waters continued to rise and fall, and then rise again to new heights. Eventually the pre-Flood mountains (which are not the same mountains that exist today) were all covered.

The sediments that were laid down one layer at a time by the ever-surging Flood waters contained organic debris. Over time, this debris either fossilized or metamorphosed (transformed) into coal, oil, and/or natural gas.

The earth's magnetic field was very dynamic during and after the Flood, resulting in more than two poles and frequent reversals of those poles. Water within the core undoubtedly aided these dynamics. With such significant changes altering the earth's protective magnetic field, it is likely that the effectiveness of the field in shielding earth from cosmic radiation was severely compromised during this turbulent time. This undoubtedly resulted in severe genetic damage to the human genome, which likely contributed to one or both of the catastrophic reductions in lifespan that are highlighted in Figure 11 on page 256.

The Flood not only destroyed all living things on the face of the land, it also destroyed the earth's equilibrium—from the inner core to the land surfaces to the atmosphere high above. Naturally, the titanic forces that had been unleashed to carry out this devastation took time to subside. Furthermore, the imbalances they brought about continued to wreak havoc as readjustments of the various natural systems gradually took place. Hundreds of years were required for equilibrium to be reestablished. Residual catastrophes took place during this time, including the Division of the earth about a century after the Flood ended.

Also, the Flood's destabilization of climate patterns, ocean temperatures, and ocean currents led to the Ice Age, which lasted until the days of Abraham and Job (approximately 300 years after the Flood). Calculations show that the ocean's excessive heat took several hundred years to dissipate, and during this time the Ice Age dominated in the northern regions of the globe.

The Flood was responsible for the sedimentary strata that contain animal fossils, since all animal life was destroyed by the Flood. These strata were deposited upon antediluvian rock by the Flood and they comprise the geologic record for the Paleozoic, Mesozoic, and Cenozoic eras (see Figure 9 on page 254).

Notice in Figure 9 that a "Cambrian Explosion" took place about 530 million years ago. This geologic and evolutionary milestone was so named because it represents the relatively rapid appearance of most marine animal life found in the fossil record—without any trace of fossilized predecessors anywhere else in the fossil record.

A nearly worldwide Precambrian-Cambrian unconformity, known as the Great Unconformity, appears in the geologic record immediately before the Cambrian Explosion. According to modern geology, an unconformity is a gap in the geological record where time appears to be missing between rock strata. Such an unconformity is identified when a layer of sediment lies upon what is considered a much older geologic layer—thus a gap in time between them is assumed to be missing. The Precambrian-

Cambrian unconformity consists of Cambrian sediments (assumed to be on the order of 525 million years old) sitting atop Precambrian rocks (assumed to be on the order of 1.5 billion years old).

As Dr. John Baumgardner explained—

"... the beginning of the Flood must correlate with the abrupt discontinuity in the fossil record at the Precambrian-Cambrian boundary. This point in the geological record not only marks a sharply-defined beginning to the abundant occurrence of fossils, but it also represents a nearly worldwide stratigraphic unconformity in the rocks themselves."

Additional evidence that the Flood was responsible for the sedimentary strata in the geologic record involves the huge areal extents of many sedimentary deposits. Geologic evidence of the destructive work of the global Flood is found on every continent.

Chapter 4 Recap

This chapter provided scientific evidence, or other basis, for the following truths about the fossil record—

- The worldwide Cambrian Explosion fossil record was created by the catastrophic Global Flood.
- Essentially no trace of transitional life forms, required for forward evolution, can be found anywhere in the fossil record. Thus, Darwin correctly pointed out a fundamental fallacy of his own theory.
- When fossils first appear in the rock strata, they are fully formed and functional.
- The evolutionary concept of punctuated equilibria has no scientific basis. Geneticists are justified in rejecting it as an evolutionary hypothesis.

- The reality of stasis (little or no evolutionary change) is manifested throughout the fossil record.
- Mainstream science no longer looks to the fossil record for evidence to support its latest theories of evolution.
- Cladograms are graphical illustrations of an individual researcher's hypotheses. They are not fact, and do not provide scientific proof of forward evolution.
- Distribution of the fossil record (95% marine invertebrates, < 5% plants, < 1% vertebrates) mirrors the distribution predicted by the Global Flood.
- The global fossil record—recorded in sedimentary strata (some of which cover hundreds of thousands of square miles)—could only have been deposited by the Global Flood.
- The typical mixture of fossil types contained within individual strata could only have been deposited by the Global Flood. The typical mixture includes terrestrial life forms, marine life forms, and vegetation from various environments.
- Fossils of animals preserved in the act of swallowing other animals, or giving live birth, provide reliable evidence that these animals were rapidly buried in a liquid grave.
- Numerous fossils of animals preserved in the act of gasping for breath provide solid evidence that these animals drowned and were rapidly buried in a liquid grave.
- The fact that scientists have discovered dinosaur mummies and other fossils harboring soft tissues, squid ink, and even blood cells cannot be denied.
- No reliable scientific explanation has been given for the existence of such mummies, soft tissues, squid ink, and blood cells in the fossil record. Logic dictates that this evidence

proves dinosaurs are not millions of years old; however, modern science will never adopt this explanation since it does not support current theories of evolution.

- Many examples of ancient art displaying images of dinosaurs have been found around the world.
- Preliminary results produced by the Neanderthal genome project show that the Neanderthal genome essentially matches that of modern humans.
- The theory that "Lucy" was able to walk upright like a human is not based on reliable scientific evidence; rather it is based on a fossilized pelvis that was deliberately modified and then pieced together so as to look like a human pelvis.
- Multiple creatures have been found living today that were thought to have been extinct for millions of years, and their bodies closely resemble those of their fossilized ancestors. Darwin called these "living fossils."
- The fossil record is essentially complete.

Significance of the Breakthrough Science Presented

The author has presented and discussed in this book breakthrough science that pertains to mortal life, the mortal earth, its geologic record, and its fossil record subsequent to the Fall of Adam and Eve some 6,000 years ago. The significant and refreshing truths that emerge from this groundbreaking science may be summarized at the highest level as follows—

1. There are no real fossils of pre-Adam transitional humans.
2. Forward evolution, as proposed by Charles Darwin and mainstream science, does not happen—ever!
3. Adaptive evolution does happen and is essential to maintain life on earth.

4. The Flood was truly global and cataclysmic in effect.
5. Catastrophic plate tectonics both caused and sustained the Flood.
6. Residual effects of the Flood brought about a single Ice Age that ended about 300 years after the Flood.
7. The Great Unconformity and the earth's sedimentary rock strata positioned above it comprise the geological record that was formed by the Flood's fury about 4,500 years ago.
8. Cambrian Explosion fossils and all other fossils located above them encompass the fossil record that was fashioned by the Flood's mercilessness roughly 4,500 years ago.
9. Most living things that contributed to the fossil record lived at the same time.
10. Dinosaurs lived both before and after the Flood.
11. Mainstream geology is based on unreliable dating methods and denies the catastrophic effects of the Global Flood.
12. Mainstream paleontology ignores the following facts: (*i*) no trace of transitional life forms can be found anywhere in the fossil record; (*ii*) the fossil record is one of stasis (showing no forward evolution); and (*iii*) AMS carbon dating results show that no dated fossils can be as old as 90,000 years. Truth number 8 (above) limits the maximum age of these fossils to approximately 4,500 years.
13. Mainstream paleontology denies that the recent discovery of dinosaur mummies, soft tissues, squid ink, and blood cells in the fossil record denotes that the accompanying fossilized animals lived within the last 4,500 years.

Many Christian readers will undoubtedly rejoice over these glorious truths since they rise above all science that offends God by refusing to confess His hand in all things.

Furthermore, the glorious truths revealed in this book resolve the uncomfortable dilemma encountered by many Christians as they study the tenets of mainstream science which whisper *there is no God who created life and the universe*. Christians can now embrace scientific truth that reinforces Biblical truth. Furthermore, the reader can see clearly why mainstream geology and paleontology are blinded to the glorious truths unveiled here. No longer will he or she be intimidated by mainstream science's foul accusation of *pseudoscience!*

Christians who do not believe that many of the tenets of mainstream science are true and are looking for real scientific confirmation of their faith in God as The Creator now have that real scientific confirmation. The scientific breakthroughs presented here both supplement and complement the Holy Bible!

Hopefully the reader's faith in God as The Creator has been increased, and the circle of truth that he or she undoubtedly cherishes has been expanded, by reading this book. And, hopefully, he or she welcomes its conclusions with joy and rejoicing.

In closing, the author leaves the reader with this counsel from an inspired Christian leader, Howard W. Hunter—

"With God our Heavenly Father, all truth, wherever found or however apprehended, is circumscribed into one great whole. Ultimately, there are no contradictions, no quarrels, no inscrutable paradoxes, no mysteries."

GLOSSARY

a priori valid independently of observation

abiogenesis (or biopoiesis) mainstream science's explanation of how life came about spontaneously from non-living, self-replicating molecules

allele a member of a gene pair that resides on a chromosome and is the mechanism by which traits are passed on from parents to offspring

asexual organisms single cell bacteria or certain types of plants and fungi that reproduce asexually

asexual reproduction offspring have only a single parent and thus inherit the genes of that parent only

asthenosphere the portion of the earth's mantle that lies directly beneath the lithosphere

axiom an established premise, principle, or rule, or a self-evident truth

base pair a pair of nucleotides that bond according to hydrogen bonding patterns; the possible base pairs are G-C (guanine-cytosine) and A-T (adenine-thymine)

biological noise any factor that diminishes the effectiveness of natural selection

Cambrian Explosion the relatively rapid appearance of most marine animal life found in the fossil record

catastrophic plate tectonics the phenomenon where the ocean lithosphere is rapidly recycled into the mantle via a runaway process

chordate features consist of a central nerve chord supported by a rod-like structure

chromosomes single pieces of coiled DNA containing many genes, DNA-bound proteins, regulatory elements, and other nucleotide sequences; they also form the "volumes" of the genetic code

cladistics an approach to biological classification where organisms are grouped together based on whether or not they have one or more shared unique characteristics that come from the group's last common ancestor and are not present in more distant ancestors

cladogram a tree-shaped diagram that is assumed to represent the best hypotheses of evolutionary relationships

continental margin a zone of the ocean floor, extending from the coast line, that includes the continental shelf, the continental slope, and the continental rise

convection the collective movement of matter within a fluid where warmer parts move up and colder parts move down; local circular or swirling flows may result

convergent evolution the independent evolution of similar features in species of different lineages

coprolite fossilized body waste material

deleterious mutation damaging or harmful

DNA deoxyribonucleic acid; a molecule that encodes the genetic instructions used in the development and functioning of all known living organisms and many viruses

dominance interaction of the alleles of a single gene pair where one allele is expressed over the other allele

entropy the measure of randomness (or disorder) of the molecules within a thermodynamic system; also the process of degradation, or degeneration, that naturally takes place with all things physical on earth

epigenetics gene changes that are not DNA changes

epistasis the interaction between genes and their modifier genes

error catastrophe extinction

eugenics the practice of improving the quality of the human race by promoting increased reproduction of people with desired traits and less reproduction of people with undesired traits

eukaryotes organisms that have both a membrane-bound cell nucleus and distinct cellular compartments; they include all forms of life except prokaryotes

family a taxonomic rank below order and above genus used to classify organisms

fenestrae natural openings in a skull

fitness ability to survive and reproduce

fitness valley the period of time during which a species remains less fit because it has a deleterious, partially-completed gene

fixed a mutation becomes permanent within a population's genome; all individuals within the population have two copies of the mutation

gamete a reproductive cell (either a sperm or an ovum cell in humans)

gastroliths fossilized rocks that were swallowed to aid digestion

genes molecular units of heredity of a living organism; they also form the "chapters" of the genetic code

genetic entropy degeneration of the genome

genus a taxonomic rank below family and above species used to classify organisms

half-life the amount of time required for a quantity to fall to half its value as measured at the beginning of the time period

heritability how likely a trait is to be passed from one generation to another; also how likely natural selection can select the trait

heritable able to be inherited

homeostasis the process which maintains the stability of an organism's internal environment in response to changes in external conditions

hybridization the ability of related species to interbreed and produce fertile offspring

igneous rock solidified magma

in situ located in something's natural place of origin

irreducible complexity the concept that a biological system composed of multiple interacting parts which contribute to its basic function will stop functioning if any one of the parts is removed from the system; also, such a system cannot arise piecemeal, but must come together all at once

junk DNA thought at one time to have no biological function

kind a biblical classification corresponding to our modern taxonomic ranks of genus and family

lahar a slurry of water, pyroclastic material, and rock fragments that are produced in a volcanic eruption

lava molten rock that is above the earth's surface (as when expelled by a volcano); also the same rock after it solidifies

lethal mutation a mutation that reduces fitness 100%—meaning there can be no reproduction of the afflicted organism

linkage blocks clusters of nucleotides that are physically linked and inherited as if a single trait—i.e., genome building blocks

Lipalian an interval of the geological clock invented by Charles Walcott to explain the period of time when ancestors of Cambrian animals evolved without leaving a fossil record; Walcott proposed that the Lipalian rocks have either been eroded or remain undiscovered

lithosphere the earth's crust and the uppermost, solid layer of the earth's mantle

magma molten rock that is below the earth's surface

mantle comprises the 1,800 mile (2,897 km) thick portion of the earth's interior directly below the earth's crust

metamorphic rock formed by extreme heat and pressure

metapopulation a group of geographically separated populations of the same species that interact, such as the human race

minor mutation a mutation that has a subtle but distinct effect on reproductive capability and resides to the left of Kimura's no selection zone; therefore it directly contributes to natural selection

modern evolutionary synthesis the integration of genetics with natural selection by Jack Haldane, Ronald Fisher, and Sewall Wright

mortality life on earth

mutation load the reduction in a population's mean fitness relative to the fitness of an idealized mutation-free individual

mutational meltdown the synergistic interaction between random genetic drift and mutation accumulation

neutral mutations have essentially a nearly-neutral (or nearly-zero) effect on an organism's fitness, and therefore are not selectable by natural selection

nuclear transmutation the transformation of an atom into a different element when its nucleus emits particles through radioactive decay

nucleosomes the basic units of DNA packaging

nucleotides molecules that make up the individual steps of the spiral staircase structure of DNA; they also form the "letters" and "words" of the genetic code

ornithischia dinosaurs have a bird hip

paleomagnetic data measurements of the magnetism found in geologic strata

Pangaea mainstream science's name for the supercontinent that existed before the Flood

peat partially decayed plant material

perturbation a deviation of a system from its normal state, caused by an outside influence

petrification organic hard part material is replaced by silica that has dissolved in percolating groundwater; the silica fills pores and cavities in the specimen through chemical precipitation

phenotype a set of observable traits

phytoplankton photosynthesizing microscopic organisms

plate tectonics large scale motions of the earth's fractured lithosphere

plumes results of the catastrophic plate tectonics runaway process; via convection, plumes of hot mantle material move upward while plumes of cooler material move downward

poly-constrained each level of meaning limits potential change

poly-functional has meaning on several different levels; e.g., a DNA sequence

polymorphism a variation of adaptive evolution that produces different sets of observable traits within the same population of a species

Primary Axiom man is merely the product of random mutations and natural selection

prokaryotes single cell organisms that lack both a membrane-bound cell nucleus and distinct cellular compartments; modern kinds include bacteria and blue-green algae

pseudoscience a theory or hypothesis without scientific foundation

punctuated equilibria a theory that, when major evolutionary change occurs, a species rapidly splits into two distinct species rather than transforming gradually into a different single species

pyroclastic flow an extremely fast-moving current of hot gas, ash, and lava (collectively called pyroclastic flow material) that can result from a volcanic eruption

pyrolized turned to char by intense heat

radiohalo (also radioactive halo) a series of concentric spheres in granite biotite mica produced by radioactive decay

random genetic drift gene frequency changes in a population (usually downward) that are determined by random occurrences rather than natural selection

ratchet a device that allows motion or function in one direction while preventing motion or function in the opposite direction

recessive requires the same gene to come from both parents in order to be expressed

recombination the process by which new combinations of genes are created within an individual to pass on to progeny; in other words, the process that ensures offspring are genetically different from each other

RNA ribonucleic acid; a ubiquitous family of large biological molecules that perform multiple vital roles in the coding, decoding, regulation, and expression of genes

saurischia dinosaurs have a lizard hip

seafloor spreading the formation of new oceanic crust along divergent boundaries of tectonic plates

Second Law of Thermodynamics a spontaneous process in a closed system will always maximize entropy, or progress toward thermodynamic equilibrium, and thus disperse energy

segregation allele pairs separate or segregate during gamete formation

selection interference selection for one trait interferes with selection for another trait

sexual dimorphism a type of polymorphism that differentiates males from females within a species in traits such as shape, structure, color, pattern, size, ornamentation, and behavior

silent DNA thought at one time to have no biological function

species a taxonomic rank below genus used to classify organisms

spreading centers places on the ocean floor where upwelling, molten mantle material (magma) rose to the ocean floor's surface during the Flood; they may have been the fountains of the great deep

stasis little or no evolutionary change

stochasticity randomness

stromatolites layered mat structures that are secreted by prokaryotes; similar to modern algae deposits

subduction zone a place where a subducting tectonic plate (normally with oceanic crust) moves beneath the other tectonic plate into the earth's mantle

synergistic epistasis several mutations cause more damage collectively than they would individually

tectonic plates massive slabs of rock, composed of continental and oceanic lithosphere, that cover the earth

tetrapods the first four-limbed vertebrates

thermal runaway instability the escalating process, resulting from rising temperature and decreasing viscosity, that causes runaway subduction of tectonic plates

transition zone the part of the earth's interior that is midway between the upper and lower boundaries of the earth's mantle

truncation selection artificial selection of individuals according to their mutation count

unconformity a break in the continental sedimentary geologic record; also a gap in the geological record where time appears to be missing between rock strata; identified when a layer of sediment lies upon what is considered a much older geologic layer

uniformitarianism the belief that what science sees happening in nature today applies to the distant past as well as the future

viscosity a liquid's "thickness" (honey has a higher viscosity than water)

References and Wikipedia Licenses

Prologue

[1] "Famous Scientists Who Believed in God" <http://www.godandscience.org/apologetics/sciencefaith.html>

Chapter 1

[1] Van Wylen, G. J. and Sonntag, R. E. *Fundamentals of Classical Thermodynamics* Second Edition. 1973. New York. John Wiley and Sons, Inc. p.248.

[2] Strahler, Arthur N. *Science and Earth History: The Evolution/Creation Controversy*. Amherst, NY. Prometheus Books. 1999. p. 62.

Chapter 2

[1] Morris, John D. *The Global Flood—Unlocking Earth's Geologic History*. Dallas, TX. Institute for Creation Research. October 2012. p. 85.

[2] Kimura, Motoo. *The Neutral Theory of Molecular Evolution*. NY, NY. Cambridge University Press. 1983. p. *ix*.

[3] Sagan, Carl. *Cosmos*. New York. Random House. 1989. p. 30.

[4] Sanford, John C. *Genetic Entropy & The Mystery of the Genome* Second Edition. Lima, NY. Elim Publishing. 2005. p. *v*.

[6] Ruse, M. Saving Darwin from the Darwinians. *National Post*. May 13, 2000. B-3.

[7] Lewontin, R. *Billions and Billions of Demons: Review of The Demon-Haunted World by Carl Sagan*. The New York Review of Books. 1997. 44(1). p 31.

[8] Haldane, J. B. S. The Cost of Natural Selection. *Journal of Genetics* Volume 55 Issue 3. December 1957. pp 511-524.

[9] Sanford, John C. *Genetic Entropy & The Mystery of the Genome* Second Edition. Lima, NY. Elim Publishing. 2005. pp 159-160.

[10] Kimura, Motoo. *The Neutral Theory of Molecular Evolution*. NY, NY. Cambridge University Press. 1983. p 27.

[11] Muller, H. J. Our Load of Mutations. *American Journal of Human Genetics* Volume 2(2). June 1950. pp 111-176.

[12] Muller, H. J. The Relation of Recombination to Mutational Advance. *Mutation Research* Volume 1 Issue 1. 1964. pp 2-9.

[13] Sanford, John C. *Genetic Entropy & The Mystery of the Genome* Second Edition. Lima, NY. Elim Publishing. 2005. pp 92-93.

[14] Kimura, Motoo. *The Neutral Theory of Molecular Evolution*. NY, NY. Cambridge University Press. 1983. pp 30-31.

[15] Sanford, John C. *Genetic Entropy & The Mystery of the Genome* Second Edition. Lima, NY. Elim Publishing. 2005. p 93.

[16] Neel, J. V. et al. The rate with which spontaneous mutation alters the electrophoretic mobility of polypeptides. *Proceedings of the National Academy of Science of the United States of America* Volume 38. January 1986. pp 389-393.

[17] Pennisi, E. ENCODE project writes eulogy for junk DNA. *Science* Volume 337 Number 6099. 7 September 2012. pp 1159-1161.

[18] Kondrashov, A. S. Contamination of the genome by very slightly deleterious mutations: why have we not died 100 times over? *Journal of Theoretical Biology* Volume 175 Issue 4. August 1995. pp 583-594.

[19] Kondrashov, A. S. Direct estimates of human per nucleotide mutation rates at 20 loci causing Mendelian diseases. *Human Mutation* Volume 21 Issue 1. January 2003. pp 12-27.

[20] Crow, J. F. Genetic Effects of Radiation. *Bulletin of Atomic Scientists* Volume 14. January 1958. pp 19-20.

[21] Crow, J. F. The high spontaneous mutation rate: is it a health risk? *Proceedings of the National Academy of Science of the United States of America* Volume 94 Number 16. 1997. pp 8380-8386.

[22] Lynch, M., Conery, J., Burger, R. Mutation accumulation and the extinction of small populations. *The American Naturalist* Volume 146 Number 4. 1995. pp 489-518.

[23] Lynch, M. and Higgins, K. Metapopulation Extinction Caused by Mutation Accumulation. *Proceedings of the National Academy of Science of the United States of America* Volume 98 Number 5. 2001. pp 2928-2933.

[24] Hoyle, F. Omni Lecture. Royal Institution. London. 12 January 1982.

[25] Hoyle, F. *Evolution from Space.* NY, NY. Touchstone Books, Simon & Schuster. 1981 pp. 27–28

[26] Hoyle, F. *Mathematics of Evolution.* Memphis, TN. Acorn Enterprises, LLC. 1999.

[27] <http://en.wikipedia.org/wiki/Level_of_support_for_evolution #cite_ref-1>

[28] Sanford, John C. *Genetic Entropy & The Mystery of the Genome* Second Edition. Lima, NY. Elim Publishing. 2005. pp *vi-viii.*

[29] Sanford, John C. *Genetic Entropy & The Mystery of the Genome* Second Edition. Lima, NY. Elim Publishing. 2005.

[30] Webster, Richard. *Why Freud Was Wrong: Sin, Science and Psychoanalysis*. Oxford. The Orwell Press. 2005. pp *ii-iii.*

[31] Kimura, M. Model of effectively neutral mutations in which selective constraint is incorporated. *Proceedings of the National Academy of Science of the United States of America* Volume 76 Number 7. July 1979. pp 3440-3444.

[32] Gerrish, P. and Lenski, R. The fate of competing beneficial mutations in an asexual population. *Genetica* 03-1998 Volume 102-103 Issue 0. pp 127-144.

[33] Bataillon, T. Estimation of spontaneous genome-wide mutation rate parameters: whither beneficial mutations? *Heredity* Volume 84 Issue 5. May 2000. pp 497-501.

[34] Nachman, M. W. and Crowell S. L. Estimate of the Mutation Rate per Nucleotide in Humans. Genetics Volume 156 Number 1. 1 September 2000. pp 297-304.

[35] Schoen, D. J. et al. Deleterious mutation accumulation and the regeneration of genetic resources. *Proceedings of the National Academy of Science of the United States of America.* Volume 95 (1). 6 January 1998. pp 394-399.

[36] Elena, S. F. and Lenski R. E. Test of synergistic interactions among deleterious mutations in bacteria. *Nature.* 390(6658). 27 November 1997. pp 395-398.

[37] Gabriel, S. B. et al. The structure of haplotype blocks in the human genome. *Science.* Volume 296 Number 5576. 21 June 2002. pp 2225-2229.

[38] Trifonov, E. N. The multiple codes of nucleotide sequences. *Bulletin of Mathematical Biology*. Volume 51 Number 4. 1989. pp 417-432.)

[39] Gibbs, W. W. The Unseen Genome: Gems among the Junk. *Scientific American*. November 2003. pp 48-53.

[40] Behe, M. *Darwin's Black Box: The Biochemical Challenge to Evolution*. The Free Press. NY, NY. 1996.

[41] Strahler, Arthur N. *Science and Earth History: The Evolution/Creation Controversy*. Amherst, NY. Prometheus Books. 1999. p 90.

[42] Ibid. p 90.

[43] <http://www.answersingenesis.org/articles/nab2/adam-and-noah-live>

[44] <http://en.wikipedia.org/wiki/Sumerian_King_List>

[45] Walton, J., The antediluvian section of the Sumerian King List and Genesis 5, *Biblical Archaeologist*, **44**:207–208, 1981.

[46] Kimura, M. How genes evolve; a population geneticist's view. Ann. Genet. Sel. Anim. Volume 19 Number 3, September 1976. pp

[47] Sanford, J.C. *Biological Information—New Perspectives, A Synopsis and Limited Commentary*. Waterloo, NY. FMS Publications. p 2.

[48] Wells, Jonathan. The Membrane Code: A Carrier of Essential Biological Information That Is Not Specified by DNA and is Inherited Apart from It. *Biological Information—New Perspectives*. World Scientific Publishing Company. July 2013.

[49] Johnson, Donald. Biocybernetics and Biosemiosis. *Biological Information—New Perspectives*. World Scientific Publishing Company. July 2013.

[50] Basener, W. F. Limits of Chaos and Progress in Evolutionary Dynamics. *Biological Information—New Perspectives*. World Scientific Publishing Company. July 2013.

[51] Montañez, G. et al. Multiple Overlapping Genetic Codes Profoundly Reduce the Probability of Beneficial Mutation. *Biological Information—New Perspectives*. World Scientific Publishing Company. July 2013.

[52] Behe, M. J. Getting There First: An Evolutionary Rate Advantage for Adaptive Loss-of-Function Mutations. *Biological Information—New Perspectives*. World Scientific Publishing Company. July 2013.

[53] Brewer, W. H. et al. Information Loss: Potential for Accelerating Natural Genetic Attenuation of RNA Viruses. *Biological Information—New Perspectives*. World Scientific Publishing Company. July 2013.

[54] McIntosh, A. C. Information and Thermodynamics in Living Systems. *Biological Information—New Perspectives*. World Scientific Publishing Company. July 2013.

[55] <http://www.evolutionnews.org/2006/02/over_500_scientists_proclaim_t001981.html>

[56] Sanford, John C. *Genetic Entropy & The Mystery of the Genome* Second Edition. Lima, NY. Elim Publishing. 2005. pp. 116-117.

Chapter 3

[1] <http://en.wikipedia.org/wiki/Neptunism>

[2] <http://en.wikipedia.org/wiki/James_Hutton>

[3] <http://en.wikipedia.org/wiki/Charles_Lyell>

[4] Strahler, Arthur N. *Science and Earth History: The Evolution/Creation Controversy*. Amherst, NY. Prometheus Books. 1999. p 62.

[5] <http://en.wikipedia.org/wiki/Geologic_time_scale>

[10] D. Russell Humphreys, Steven A. Austin, John R. Baumgardner, and Andrew A. Snelling. "Helium Diffusion Age of 6,000 Years Supports Accelerated Nuclear Decay." June 2004. *Creation Research Quarterly Journal*. Volume 41 Number 1.

[11] Morris, John D. *The Global Flood—Unlocking Earth's Geologic History*. Dallas, TX. Institute for Creation Research. October 2012. p 103.

[12] G. B. Dalrymple, *The Age of the Earth*. 1991. Stanford, CA, Stanford University Press. p. 91.

[13] Snelling, A. A. 1999. "Excess Argon": The "Achilles' Heel" of Potassium-Argon and Argon-Argon "Dating" of Volcanic Rocks. *Acts & Facts*. 28 (1).

[14] Baumgardner, J. 2003. Carbon Dating Undercuts Evolution's Long Ages. *Acts & Facts*. 32 (10).

[15] R. Baumgardner, D. R. Humphreys, A. A. Snelling, and S. A. Austin. "Measurable ^{14}C in fossilized organic materials: Confirming the young earth creation/Flood model." 2003. *Proceedings of the Fifth International Conference on Creationism*. R. E. Walsh, Editor. Creation Science Fellowship. Pittsburgh, PA. pp. 127-142.

[16] F. H. Schmidt, D. R. Balsley, and D. D. Leach. 1987. "Early expectations of AMS: Greater ages and tiny fractions. One failure?—one success." *Nuclear Instruments and Methods in Physics Research B*. 29:97-99.

[17] Giem, P. 2001. Carbon-14 content of fossil carbon. *Origins*. 51, 6–30. Also see <http://www.grisda.org/origins/51006.htm>

[18] Marie-Josée Nadeau1, Pieter M. Grootes, Antje Voelker, Frank Bruhn, Alexander Duhr, Angelika Oriwall. 2001. Carbonate 14C Background: does it have multiple personalities? Leibniz-Laboratory. Radiocarbon 43 (2A), pp. 169–176.

[19] Baumgardner, J. 2003. Carbon Dating Undercuts Evolution's Long Ages. *Acts & Facts*. 32 (10).

[20] Polonium Radiohalos by Laurence D Smart. <http://unmaskingevolution.com/12-radiohalos.htm>

[21] Andrew A. Snelling, Ph.D. 2000. Polonium Radiohalos: Still "A Very Tiny Mystery". *Acts & Facts*. 29 (8).

[22] Robert V. Gentry. 1986. *Creation's Tiny Mystery*. Knoxville, TN. Earth Science Associates. p. 66.

[23] <http://en.wikipedia.org/wiki/Robert_V._Gentry>

[24] Humphreys, D. R. The Mystery of Earth's Magnetic Field. 1989. Acts & Facts 18 (2).

[25] Morris, J. D. *The Global Flood*. Dallas, TX. October 2012. Institute for Creation Research. p 37.

[26] <http://en.wikipedia.org/wiki/Dynamo_theory>

[27] D. Russell Humphreys. Reversals of the Earth's Magnetic Field During the Genesis Flood. Presented at the First International Conference on Creationism, Pittsburgh, PA. August 4–9, 1986. Published in: Proceedings of the First International Conference on Creationism. 1986. R. E. Walsh, C. L. Brooks, and R. S. Crowell (Editors). pp. 113–123.

[28] D. Russell Humphreys. Physical Mechanism for Reversals of the Earth's Magnetic Field During the Flood. Presented at the Second International Conference on Creationism, Pittsburgh, PA. July 30–August 4, 1990. Published in: Proceedings of the Second International Conference on Creationism. 1990. R. E. Walsh and C. L. Brooks (Editors). pp. 129–140.

[29] Brian Thomas. Mercury's Fading Magnetic Field Fits Creation Model. <http://www.icr.org/article/mercurys-fading-magnetic-field-fits/>

[30] D. Russell Humphreys. The Creation of Cosmic Magnetic Fields. In 2008. A. A. Snelling (Editor). Proceedings of the Sixth International Conference on Creationism. pp. 213–230. Pittsburgh, PA: Creation Science Fellowship and Dallas, TX: Institute for Creation Research.

[31] Thomas, B. Continents Should Have Eroded Long Ago. 22 August 2011. Institute for Creation Research. <http://www.icr.org/article/continents-should-have-eroded-long/>

[32] Portenga, E. W. and R. R. Bierman. 2011. Understanding Earth's eroding surface with ^{10}Be. *GSA Today*. 21 (8): pp. 4-10.

[33] Roth, A. A. 1986. Some Questions about Geochronology. *Origins*. 13 (2): pp. 64-85. <http://www.grisda.org/origins/13064.htm>

[35] <http://en.wikipedia.org/wiki/Pangaea>

[36] Morris, J. D. *The Global Flood*. Dallas, TX. October 2012. Institute for Creation Research.

[37] Ibid. p. 74.

[38] Ibid. p 45.

[40] Ibid. p 79.

[41] Ibid. p. 77.

[42] <http://en.wikipedia.org/wiki/Antonio_Snider-Pellegrini>

[43] <http://en.wikipedia.org/wiki/Plate_tectonics>

[44] Austin, Stephen A.; Baumgardner, J.R.; Humphreys, R.D.; Snelling, A.A.; Vardiman, L.; Wise, K.P. 1994. "Catastrophic Plate Tectonics: A Global Flood Model of Earth History". Third International Conference on Creationism. Pittsburgh, PA. July 18–23, 1994: Institute for Creation Research.

[45] <http://en.wikipedia.org/wiki/ Continental margin>

[46] Dziewonski, A. M. 1984. Mapping the lower mantle: Determination of lateral heterogeneity in P velocity up to degree and order 6. *Journal of Geophysical Research* 89: pp. 5929–5952.

[47] Engebretson, D. C., Kelley, K. P., Cushman, H. J., and Reynolds, M. A. 1992. 180 million years of subduction. *GSA Today* 2(5). pp. 93–100.

[48] John R. Baumgardner. Catastrophic Plate Tectonics: The Physics Behind the Genesis Flood. Fifth International Conference on Creationism. Pittsburgh, PA. August 4–8, 2003: Creation Science Fellowship, Inc.

[49] Strom, R.G., Schaber, G.G., and Dawson, D.D. 1994. The Global Resurfacing of Venus. *Journal of Geophysical Research* Volume 99 pp. 10899-10926.

[50] Morris, J. D. *The Global Flood*. Dallas, TX. October 2012. Institute for Creation Research. p. 37.

[51] <http://en.wikipedia.org/wiki/Mount_Ararat>

[52] <http://en.wikipedia.org/wiki/Judi_Dagh>

[53] <http://en.wikipedia.org/wiki/Mount_Ararat>

[54] Austin, S. A. 1979. Depositional environment of the Kentucky No. 12 coal bed (Middle Pennsylvanian) of western Kentucky, with special reference to the origin of coal lithotypes. Unpublished doctoral dissertation. Pennsylvania State University.

[55] Morris, J. D. *The Global Flood*. Dallas, TX. October 2012. Institute for Creation Research. p. 63.

[56] <http://en.wikipedia.org/wiki/1980_eruption_of_Mount_St._Helens>

[57] <http://en.wikipedia.org/wiki/Spirit_Lake_(Washington)>

[58] Austin, S. A. 1986. Mt. St. Helens and Catastrophism. *Acts & Facts*. 15 (7).

[59] <http://www.evolutionnews.org/2013/08/does_the_great_075501.html>

[60] <http://en.wikipedia.org/wiki/Cambrian_explosion>

[61] <http://www.britannica.com/EBchecked/topic/342777/Lipalian-interval>

[62] <http://www.evolutionnews.org/2013/08/does_the_great_075501.html>

[63] Peters, S. E. and Gaines, R. R. Formation of the 'Great Unconformity' as a trigger for the Cambrian explosion. 19 April 2012. *Nature* 484, pp. 363–366.

[64] <http://www.ia.ucsb.edu/pa/display. aspx?pkey=2364>

[65] Baumgardner, John. Numerical Simulation of the Large-Scale Tectonic Changes Accompanying the Flood. 1986. *Proceedings of the First International Conference on Creationism*, R. E. Walsh, C. L. Brooks, & R. S. Crowell (Eds.), pp. 17–28.

[66] Morris, J. D. *The Global Flood*. Dallas, TX. October 2012. Institute for Creation Research. pp. 147-148.

[67] Barnhart, W. R. A Hydrodynamic Interpretation of the Tapeats Sandstone Part I: Basal Tapeats. Spring 2012. *Creation Research Society Quarterly* Volume 48 No. 4 pp. 288-311.

[68] Morris, J. D. *The Global Flood*. Dallas, TX. October 2012. Institute for Creation Research. pp. 111-112.

[69] Ager, D.V. 1973. *The Nature of the Stratigraphical Record*. London. MacMillan. pp 4-6.

[70] Morris, J. 1998. Doesn't Carbon Dating Prove The Earth Is Old? *Acts & Facts*. 27 (6).

[71] Saxena, S. K.; Dubrovinsky, L.; Rekhi, S.; Wang, Z.; Shen, G. Oceans of Water in the Earth's core. American Geophysical Union. Spring Meeting 2002. Abstract #M42A-04.

[72] D. G. Pearson, F. E. Brenker, F. Nestola, J. McNeill, L. Nasdala, M. T. Hutchison, S. Matveev, K. Mather, G. Silversmit, S. Schmitz, B. Vekemans, L. Vincze. Hydrous mantle transition zone indicated by ringwoodite included within diamond. *Nature*, 2014; 507 (7491): 221.

[73] Scientists Suspect Oceans of Water Deep Inside the Earth. <http://www.topsecretwriters.com/2014/03/scientists-suspect-oceans-of-water-deep-inside-the-earth/>

[74] University of Alberta. "Water-rich gem points to vast 'oceans' beneath Earth's surface, study suggests." ScienceDaily. ScienceDaily, 12 March 2014. <www.sciencedaily.com/releases/2014/03/140312150229.htm>

[75] <http://www.livescience.com/1312-huge-ocean-discovered-earth.html>

[76] <http://www.npr.org/blogs/thetwoway/2014/06/15/322246690/deep-below-oceans-of-water-may-be-trapped-in-a-crystal-sponge>

[77] Brandon Schmandt et al. Dehydration melting at the top of the lower mantle. *Science* 13 June 2014: vol. 344 no. 6189 pp. 1265-1268.

[78] R. H. Brown. The Interpretation of C-14 dates. *Origins,* 1979; Volume 6 number 1: pp. 30-44.

[79] H. W. Scharpenseel, P. Becker-Heidmann. Twenty-Five Years of Radiocarbon Dating Soils: Paradigm of Erring and Learning. *Radiocarbon,* 1992; 34 (3): pp. 541-549.

Chapter 4

[1] Eldredge, Niles and S. J. Gould. 1972. Punctuated equilibria: an alternative to phyletic gradualism. in *Models in Paleobiology*. Edited by Schopf. San Francisco: TJM Freeman Cooper. pp. 82-115.

[2] Gould, S. J. Evolution's Erratic Pace. May 1977. *Natural History*. Volume 86 Issue 5. pp 12-16.

[3] Gould, S. J. May 1977. *Natural History*. Volume 86 Issue 5. pp 22-30.

[4] Stanley, S. *Macroevolution*. 1977. San Francisco, CA. Freeman. p. 39.

[5] Ridley, M. Who doubts evolution? 5 June 1981. *New Scientist*. 90:832.

[6] Lewontin, R. *Billions and Billions of Demons: Review of The Demon-Haunted World by Carl Sagan*. The New York Review of Books. 1997. 44(1). p 31.

[7] Koonin, E. V. The Biological Big Bang model for the major transitions in evolution. 2007. *Biology Direct*. 2:21.

[8] <http://en.wikipedia.org/wiki/Cladistics>

[9] Morris, J. D. and Sherwin, F. J. *The Fossil Record—Unearthing Nature's History of Life*. Dallas, TX. October 2011. Institute for Creation Research. pp. 62-63.

[10] Adrain, Jonathan M., Edgecombe, Gregory D. and Lieberman, Bruce S. 2002. *Fossils, Phylogeny, and Form: An Analytical Approach*. New York: Kluwer Academic.

[11] Morris, J. D. and Sherwin, F. J. *The Fossil Record—Unearthing Nature's History of Life*. Dallas, TX. October 2011. Institute for Creation Research.

[12] Morris, J. D. and Sherwin, F. J. *The Fossil Record—Unearthing Nature's History of Life*. Dallas, TX. October 2011. Institute for Creation Research. pp. 39-40.

[13] Ibid. p. 123.

[14] Ibid. p. 43-44.

[16] <http://en.wikipedia.org/wiki/Prokaryote>

[18] <http://en.wikipedia.org/wiki/Eukaryotes>

[19] Morris, J. D. and Sherwin, F. J. *The Fossil Record—Unearthing Nature's History of Life*. Dallas, TX. October 2011. Institute for Creation Research. pp. 43-44.

[21] <http://en.wikipedia.org/wiki/Burgess_Shale>

[22] Strahler, Arthur N. *Science and Earth History: The Evolution/Creation Controversy*. Amherst, NY. Prometheus Books. 1999. p 408.

[23] <http://en.wikipedia.org/wiki/Tiktaalik>

[24] Fossil footprints give land vertebrates a much longer history, Uppsala University press release. Jan 7, 2010 reporting research published in G. Niedzwiedzki, P. Szrek, K. Narkiewicz, M. Narkiewica, and P. E. Ahlberg. 2010. Tetrapod trackways from the early Middle Devonian period of Poland Nature, 463 (7277): pp. 43-48.

[25] Ruse, M. Saving Darwin from the Darwinians. *National Post*. May 13, 2000. B-3.

[26] Stahl, B. 1985, *Vertebrate History: Problems in Evolution*. New York: Dover Publications, Inc. pp. 268 and 276.

[27] C. P. Hickman, Jr., L. S. Roberts, and A. Larson. 1997. *Integrated Principles of Zoology*. New York. McGraw-Hill. p. 549.

[28] <http://en.wikipedia.org/wiki/Ichthyosaur>

[29] <http://en.wikipedia.org/wiki/Convergent_evolution>

[30] Colbert, E. H., M. Morales, and E. C. Minkoff. 2001. *Colbert's Evolution of the Vertebrates: A history of the back-boned animals through time*. 5th edition. New York: Wiley-Liss, Inc. p. 159.

[31] Ibid. p. 128.

[32] Morris, J. D. and Sherwin, F. J. *The Fossil Record—Unearthing Nature's History of Life*. Dallas, TX. October 2011. Institute for Creation Research. p. 75.

[33] Benton, M. J. 2005. *Vertebrate Palaeontology*. Malden, MA: Blackwell Science. p 244.

[34] Colbert, E. H., M. Morales, and E. C. Minkoff. 2001. *Colbert's Evolution of the Vertebrates: A history of the back-boned animals through time*. 5th edition. New York: Wiley-Liss, Inc. p. 154.

[35] Stahl, B. 1985, *Vertebrate History: Problems in Evolution*. New York: Dover Publications, Inc. p. 336.

[36] Schweitzer, M. H., J. L. Wittmeyer, J. R. Horner, and J. K. Toporski. 2005 March 25. Soft-Tissue Vessels and Cellular Preservation in Tyrannosaurus rex. *Science*. 307 (5717): 1952-1955.

[37] June 2, 2005 News Release from North Carolina State University titled "Dinosaur Fossil Bone Leads to Gender, Age Determinations."
<http://www.ncsu.edu/news/press_releases/05_06/133.htm>

[38] H. Fields. May 2006. Dinosaur Shocker. *Smithsonian Magazine*. pp. 50-55.

[39] R. F. Service. 1 May 2009. 'Protein' in 80-Million-Year-Old Fossil Bolsters Controversial *T. Rex* Claim. *Science* Volume 324 (Number 5927): p. 578.

[40] Schweitzer, M. H. et al. 1 May 2009. Biomolecular Characterization and Protein Sequences of the Campanian

Hadrosaur *B. canadensis. Science* Volume 324 (Number 5927): pp. 626-631.

[41] Morris, J. D. *The Global Flood*. Dallas, TX. October 2012. Institute for Creation Research. p. 126.

[42] <http://www.telegraph.co.uk/science/science-news/5794280/Scientists-draw-squid-using-its-150-million-year-old-fossilised-ink.html>

[43] <http://www.dailymail.co.uk/sciencetech/article-1207367/The-150million-year-old-squid-fossil-perfectly-preserved-scientists-make-ink-ink-sac.html>

[44] <http://www.mummydinosaur.com/index.html>

[45] Schweitzer. M.H. et al. A role for iron and oxygen chemistry in preserving soft tissues, cells and molecules from deep time. *Proceedings of the Royal Society B*. 22 January 2014. Volume 281 Number 1775 20132741.

[46] <http://news.ncsu.edu/releases/schweitzer-iron/>

[47] <http://www.icr.org/article/dinosaur-soft-tissue-preserved-by-blood/>

[48] Morris, J. 2008. The Dinosaur Next Door. *Acts & Facts*. 37 (6): 13.

[49] Thomas, B. 24 June, 2013. Jungle-Covered Ruins May Hold Surprising Hints. <www.icr.org/articles/view/7546/368/>

[50] Brian Thomas, M.S. 2014. Amber Fossils Redefine 'Age of Reptiles'. *Acts & Facts*. 43 (3).

[51] Darwin, C. 1872. *The Origin of Species*, 6th edition. New York, NY: Collier Books, p. 172.

[52] <http://en.wikipedia.org/wiki/Archaeopteryx>

[53] Morris, J. D. and Sherwin, F. J. *The Fossil Record—Unearthing Nature's History of Life*. Dallas, TX. October 2011. Institute for Creation Research. p. 86.

[54] Ibid. p. 93.

[55] Ibid. p. 105.

[56] Ibid. p. 106.

[57] <http://en.wikipedia.org/wiki/Neanderthal_genome_project>

[58] <http://en.wikipedia.org/wiki/Base_pair>

[59] Jeffrey Tomkins, Ph.D. 2014. Ancient Human DNA: Neandertals and Denisovans. *Acts & Facts*. 43 (3).

[60] Tomkins, J. New Research Evaluating Similarities Between Human and Chimpanzee DNA. 2013. Proceedings of the Seventh International Conference on Creationism, Pittsburgh, PA: Christian Science Fellowship.

[61] Morris, J. D. and Sherwin, F. J. *The Fossil Record—Unearthing Nature's History of Life*. Dallas, TX. October 2011. Institute for Creation Research. p. 109.

[62] Ibid. p. 110.

[63] Ibid. p. 83.

[64] Ibid. p. 94.

[65] <http://www.home-school.com/Articles/hominids-hoaxes.html>

[66] Smoot, S. Edgar. *Lost American Antiquities*. New York. October 2013. Legends Library. pp. 126,128.

Wikipedia Licenses and Fair Use Rationale

Certain reproductions of images used in this book were obtained from the English *Wikipedia* website (*Wikipedia, The Free Encyclopedia* at www.wikipedia.com) via the *Wikimedia Commons* (see http://commons.wikimedia.org/wiki/Main_Page). Credit for authors is given where authors are known. Some reproductions are licensed under the *GNU Free Documentation License* and/or the *Creative Commons Attribution-Share-Alike License*. Others are in the U.S. public domain, as noted under the image.

Any of the images licensed under the *GNU Free Documentation License* and/or the *Creative Commons Attribution-Share-Alike License* may be reproduced under the conditions of their respective licenses, which are documented in the web addresses given below--

- <http://en.wikipedia.org/wiki/Wikipedia:Text_of_the_GNU_Free_Documentation_License>
- <http://creativecommons.org/licenses/by-sa/3.0/>

Fair use rationale for including these images is that their primary use is for informational and educational purposes; also that no free equivalents are available or could be created that would adequately give the same information. Reproduction of the images is not believed to limit the copyright owners' rights or profit in any way.

INDEX

About the Author

Upon graduating with a Bachelor of Science degree, Gary T. Wright was named the "Outstanding Student Engineer" at the University of Utah. While subsequently working for IBM as a development engineer, he earned a Master of Science degree at the University of Colorado and was awarded US Patent 4003091.

Gary was also a lead engineer at TerraTek's Drilling Research Laboratory and he taught Fluid Mechanics at the University of Utah. Later on, he became a Systems Engineer with IBM and then moved into executive management with various computer companies, including a Vice President assignment with Hewlett-Packard. Most recently, he has been in a leadership role with the emerging digital preservation industry.

Throughout his career, Gary has written numerous technical articles for various publications, and is a published author of three historical books. He is also a composer of classical music and has performed with several professional symphony orchestras as a violinist.

Gary's lifelong passion has been to uncover scientific truths that confirm the scriptures.